Steel and Grit

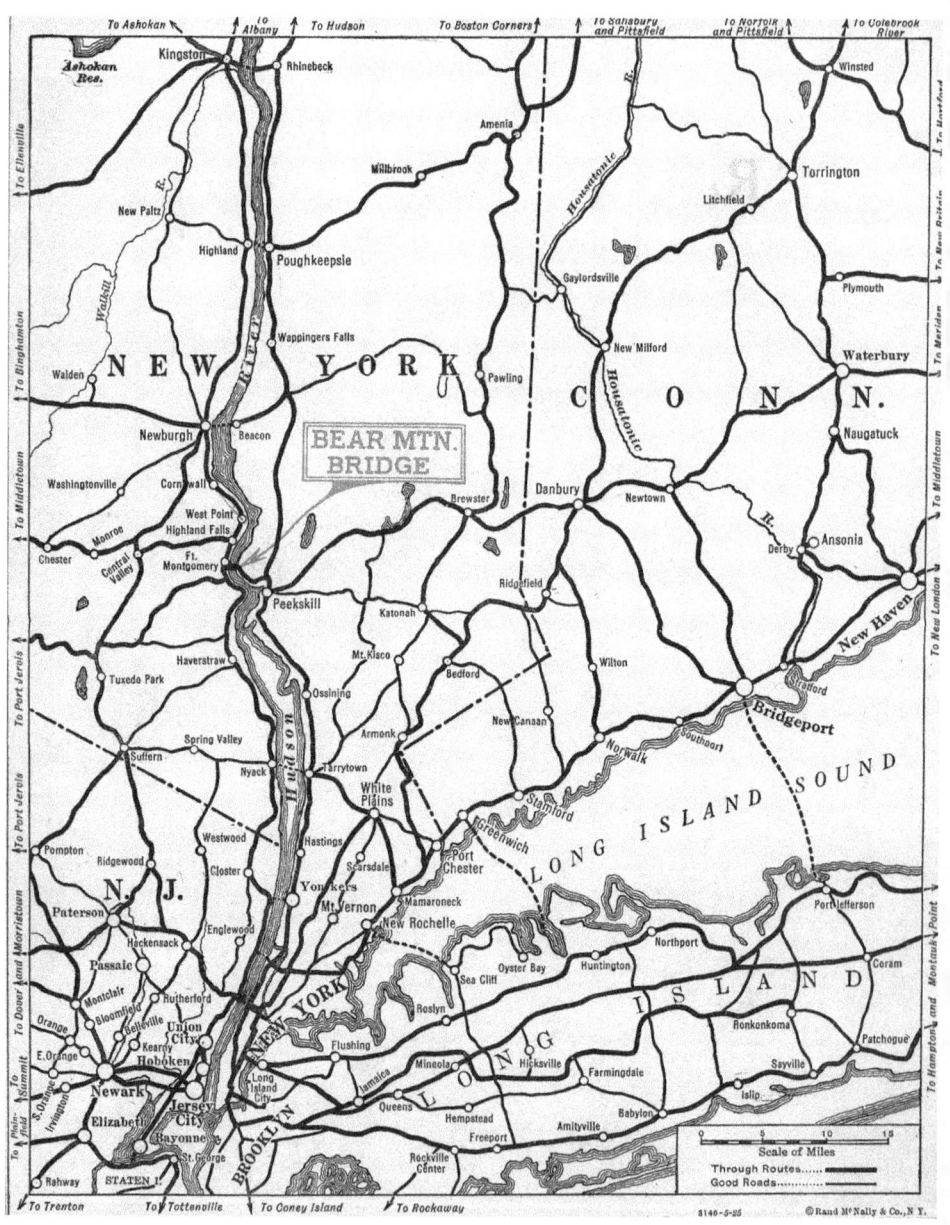

1. Map of the lower Hudson Valley, pinpointing the location of the Bear Mountain Bridge. Courtesy of the Roebling Museum; *Construction of Parallel Wire Cables for Suspension Bridges.*

STEEL & GRIT

A History of the Lower Hudson Valley and the Bear Mountain Bridge

Barbara Hansen Cali

Syracuse University Press

Copyright © 2026 by Syracuse University Press
Syracuse, New York 13244-5290

All Rights Reserved

First Edition 2026

26 27 28 29 30 31   6 5 4 3 2 1

For a listing of books published and distributed by Syracuse University Press, visit https://press.syr.edu.

ISBN: 9780815611974 (paperback)
         9780815657569 (e-book)

Library of Congress Cataloging in Publication Control Number: 2025025600

The authorized representative in the EU for product safety and compliance is Mare Nostrum Group B.V. Mauritskade 21D, 1091 GC Amsterdam, The Netherlands
gpsr@mare-nostrum.co.uk

Publication supported by a grant from

The Community Foundation *for* Greater New Haven

as part of the Urban Haven Project

I dedicate this book to John and Louise Hansen, my parents, who instilled in me a love of reading, along with curiosity and respect for history and those who came before us

Contents

Illustrations *xi*
Acknowledgments *xiii*

Introduction *1*
1. You Are Cordially Invited to the Opening *8*
2. The Valley *18*
3. Wall Street and Railroads *41*
4. The Right Man for the Job *52*
5. The Original Jersey Girls *66*
6. A Perfect Candidate *76*
7. A Partnership Is Formed *84*
8. Progress and Development *92*
9. The Devil's Horse Race *96*
10. The Chief Engineer *105*
11. Like Father, Like Son *115*
12. We Must Not Forget *122*
13. A Man with a Plan *128*
14. Baird *134*
15. Bear Mountain Hudson River Bridge Company *139*
16. Terry & Tench *147*
17. A Difference of Opinion *159*
18. Construction Begins *171*
19. Innovations *207*
20. The Bond *221*
21. Remembering *227*
22. Update—1927 through the Present *240*
Epilogue *246*

Glossary *253*
Notes *257*
Bibliography *277*
Index *283*

Illustrations

1. Map of the lower Hudson Valley, pinpointing the location of the Bear Mountain Bridge   *frontispiece*
2. West Point marching band leading parade of cars across the Bear Mountain Bridge on opening day, 1924   7
3. E. H. and Mary Harriman upon their return from Europe, 1909   36
4. Portrait of George W. Perkins Sr.   37
5. William Addams Welch at work at the Palisades Interstate Park Commission   38
6. Young W. Averell Harriman presenting a check for $1 million to George W. Perkins of the Palisades Interstate Park Commission on behalf of the Harriman family, 1910   39
7. Handwritten note from Katharine Sauzade to Cecilia G. Holland, 1901   40
8. Ironworkers who built the Bear Mountain Bridge on the bridge on opening day, 1924   60
9. Portrait of George W. Perkins Jr., member and president of the Palisades Interstate Park Commission, an investor in the Bear Mountain Bridge, and a member of the Bear Mountain Hudson River Bridge Company   61
10. Portrait of Howard Carter Baird, designer of the Bear Mountain Bridge   62
11. Frederick Tench, contractor (with Emma Tench), at Tench family gathering   63
12. Portrait of E. Roland Harriman   64
13. E. Roland Harriman and Mary Harriman at opening-day ceremonies for the Bear Mountain Bridge   65
14. Plate 2, excavation of the western anchorage pits   192
15. Plate 4, tower nearing completion   193
16. Plate 7, construction of the footbridges   194
17. Plate 10, men spinning cables   195

18. Plate 15, adjusting cables in anchorage   196
19.1. Plate 21, looking east, wheels carrying cables   197
19.2. Plate 22, looking east, wheels carrying cables   197
20. Plate 32, road construction of the east-approach highway   198
21. Plate 37, squeezing gang compacting cable   199
22. Plate 43, western anchorage   200
23. Plate 49, suspenders in place, trusses under way   201
24. Plate 51, placing the final truss connecting the east and west portions of the bridge   202
25. Plate 53, correct cable construction   203
26.1. Plate 63, handrail ropes, east anchorage   204
26.2. Plate 64, handrail ropes, east anchorage   204
27. Plate 67, commemorative plaque on the bridge on opening day   205
28. Plate 68, the finished bridge   206
29. Bear Mountain Bridge, circa 1930s   236
30. Women's Memorial Tower   237
31. Program cover for the Women's Memorial dedication   238
32. Program pages for the Women's Memorial dedication   239
33. Diagram of a suspension bridge with various parts identified   253

Acknowledgments

As a first-time author, I began a journey for which I had no road map. I have been very fortunate to have encountered so many helpful and supportive people along the way who guided me in my search for information, and I would like to express my heartfelt gratitude to all.

First and foremost, my thanks to Michelle Figliomeni, president of the Orange County Historical Society (OCHS) in Arden, New York. Michelle is also the town historian for Goshen, New York. She was the first person I encountered when I began this journey, and she not only guided me in my search for information but has been kind and gracious in her support of this project in every way. I also want to thank Donald Bayne and Grant Miller, trustees of the OCHS, and the entire staff at the OCHS for their help and support.

Besides the OCHS, I made numerous trips to the New York Public Library, the main branch and the SIBL Branch (Science, Industry & Business Library). My thanks to all the librarians and staff members who helped me access materials regarding the bridge and the people involved. On multiple occasions, I visited the Rare Books & Manuscripts Library, Butler Library, at Columbia University, where I was able to access the personal papers of both George W. Perkins Sr. and George W. Perkins Jr. Thank you to the helpful librarians and staff there. Also, thanks to Kim Stucko of Fields Library in Peekskill, New York; Frank Goderre, tollhouse historian, Bear Mountain Bridge, town of Cortlandt; and Susanne Pandich and Patrick Raftery of the Westchester Historical Society. Also, thanks to the Warner Library in Tarrytown, New York, and the staff of the New York State Archives at the New York State Library in Albany; those individuals at the Historical Society of Pennsylvania in Philadelphia and the staff and historians at the Phoenixville Historical Society in Phoenixville, Pennsylvania; and those persons who assisted me at the Hagley Museum in Wilmington, Delaware. Thanks also to Fernanda H. Perrone, with the Special Collections and University Archives at Rutgers State University in New Brunswick, New Jersey; Kathryn Burke, director, Historic Bridges of the Hudson Valley, Highland, New York; and Tara Sullivan, New York State Bridge Authority, acting executive director, retired. I am also grateful to Varissa McMickens Blair, the former executive director of the Roebling Museum in Roebling, New Jersey, along with Lynne Calamia, Roebling's current executive director. I want to express a special thanks to Matthew Shook, director of development and special projects at the Palisades Interstate Park Commission, Bear Mountain, New York, and his staff, for allowing me access to the PIPC historical archives at Iona Island numerous times. These opportunities yielded a treasure trove of information. I want to thank Erick Nielson at the Palisades Interstate Park Commission's New Jersey office, in Alpine, for his assistance as well.

A special thanks goes to the granddaughter of Frederick Tench, Margaret Tench Hamilton Crothers, and her family, without whose help and contribution I would not have been able to give Frederick Tench his due. Margaret and I exchanged many emails in 2018 when I first began my research, and I thoroughly enjoyed our correspondence. Sadly, Margaret passed away on January 17, 2021. I had hopes of meeting her in person at some point once the book was finished. She was proud of her grandfather and eager to provide me with an understanding of his personality. Although she did review an early draft of a chapter on Frederick, I wish she had had the opportunity to read the entire manuscript. Also, my thanks go out to Wendy Welch and her sister, Heather, granddaughters of William A. Welch, for their help and contribution in bringing Major Welch to life and in giving him the credit he so deserves for his part in preserving the Hudson Valley, as well as his contributions to our current system of national parks. I was fortunate to have had the opportunity to interact with these women, and I hope the result was a clearer, more interesting glimpse into the lives of both these great men of history whom they called Grandfather.

A special thanks to Adam Rosen, my editor, who provided me with guidance and direction through his corrections and reviews of my work.

He lent his educated eye to my manuscript and helped me format and detail my research material, resulting in a more appealing and well-organized text for my reader.

And, most important, to my family and friends who have been behind me 100 percent. To Carin, my daughter, who not only supported and encouraged me, but also acted as a research assistant and an IT troubleshooter whenever my computer acted up. To John Lishak, my partner and best friend, who encouraged me and read many of the preliminary chapters with an enthusiasm that allowed me to think I could do this work. To Judy, my sister, and Carol B. Allen, my friend and a fellow author, for their support and encouragement. And to Milli, my dear friend, who acted as a reader early on and expressed such an interest in the story. All of you played an important part in my ability to keep going. I want to also mention Larry Goldberg, a friend who worked professionally as a bridge engineer for many years and lent his educated eye to my writing by reviewing several chapters. Thank you all.

My faith in God and my firm belief that He directs us in the paths we take have played a big part in this achievement. The support and encouragement of all the wonderful people mentioned above have helped me along my journey. It's been one of my life's most challenging and, at the same time, most rewarding experiences. Thank you to each of you for the part you played!

Steel and Grit

Introduction

> Bridges may connect destinations but [they] also join the souls of man.
> —Author unknown

For as long as I can remember, bridges have enchanted me. However, when I share that fact with others, I inevitably receive a silent and somewhat quizzical stare. Most bridges have an industrial appearance, usually characterized by open-frame steel trusses, and many people regard them as anything but enchanting. Bridges are considered functional structures that serve a purpose. They allow us to cross from one side of a natural impediment to the other—that's it.

But I see them differently. When I come upon a bridge, something stirs within me: the visual symmetry pleases my artistic eye, while my intellectual curiosity for the story behind the monument jumps forth. They present an image of power and a sense of majesty. The sweeping catenary of the cables strung from one tower to the next is almost regal, and the eye of the beholder appreciates the order and balance as one's perspective adjusts. At the same time, they are a symbol of humanity, monuments of metal and stone joining people and landscapes, each with its personality and story. Each spans history, and through its story of circumstances and people particular to its era, it connects us to those individuals who came before. Often there were obstacles—political, financial, or otherwise—that made the quest pursued by its champions that much more challenging.

The story of how all these elements played out is often inspiring and always interesting. All bridges had at least one (sometimes more) designer or engineer who studied the natural conditions of the selected site and created a structure specifically suited to its unique location and function.

Each was built by a crew of skilled and unskilled workers with human hands that sweated or froze (possibly both) in adverse weather conditions, often risking life and limb in the process. Some, I imagined, might even have been killed during construction. I wanted to know who these people were. I wanted to tell their stories.

I have lived all my life in the greater New York–New Jersey metropolitan area and have driven or walked over the many bridges that are a part of our landscape. I have also passed under them by boat, traveling the Hudson and East Rivers and their adjoining waterways. In fact, a boating trip that took me up the Hudson on a weeklong journey several years ago inspired this book. I found beauty, strength, and majesty not only in the Hudson River itself but in each bridge we passed under.

I selected the Bear Mountain Bridge (BMB) as my focus because of its refined contours and the beauty of its location. The crossing between Anthony's Nose and Bear Mountain is one of the most picturesque locations along the river, and it most definitely adds to the overall appearance and impression of the bridge. The sleek, tall, tapering towers provide a certain elegance, and the catenary curve of the main cables adds grace and nobility. Far from emitting a sense of clumsiness and frumpery (as was noted by early critics), the bridge rather transmits refined sophistication. Its image and personality permeate the area and beckon the observer to come and learn more about this monument that has survived the past hundred years. The history of the bridge is so interwoven with the history of the Hudson Valley and with the people who lived there at that time and loved the valley enough to dedicate their lives to preserving and developing it. There was an evolving technology at the time that seemed to promise prosperity, but, at the same time, it threatened to change the very essence of the land they loved. The dedication and the ingenuity of these people fascinated me, and I realized that they—the people—are the real story. The Bear Mountain Bridge is truly a towering tribute to those individuals who recognized the importance of not only building it but also protecting the legacy of the Hudson River valley where it stands, preserving it for all of us who would come after them.

There were many involved: Frederick Tench, the contractor who, in 1921, decided that a bridge across the Hudson should be built at Bear

Mountain and was determined to be the contractor who accomplished this feat; George W. Perkins Sr., who, at the turn of the twentieth century, long before the bridge was even thought of, worked to stop the destruction of the Palisades and as a result was involved in creating the Palisades Interstate Park Commission and the valley's park system that provides the perfect setting; and the ladies of the New Jersey State Federation of Women's Clubs (NJSFWC), without whose passion and commitment the present-day Hudson Valley may have taken on an entirely different quality of existence. These individuals and many others are a real part of this story, but they were unknown to me until I was knee-deep in research. Their efforts were a key part of the formation of the PIPC that would preserve the valley and have a significant role as its protector and developer. One of the first accomplishments of the PIPC was the formation of the Palisades Interstate Park. The park was a destination for many citizens over the years, and in the early 1920s, easy access to it became one of the incentives that instigated the need for a bridge. Several members of the Harriman family, one of New York's most prominent dynasties, were significant players in saving the valley and building the bridge. As heirs to a railroad empire built by Edward H. Harriman, its patriarch and one of the country's early railroad tycoons, the family members had access to the means and the connections necessary, and they continued his legacy by supporting conservation through philanthropy and community service. The story of how E. H. Harriman came to purchase the land where he would establish his family estate and then eventually gift that land, parcel by parcel, to the state of New York is a captivating part of the valley's historical fabric and evolution. The narrative continues as we chronicle the Harrimans' intricate involvement in building the bridge.

Like many historical developments, the establishment of the PIPC and the building of the bridge were responses to the circumstances and values of the time. The era of the late nineteenth and early twentieth centuries was a fascinating period in American history. Innovations developed by creative thinkers, unencumbered by today's regulations and restrictions, were numerous. The transition from manual to mechanical gave rise to enthusiasm along with a spirit of collaboration, often exposing a patriotic spirit among citizens building a nation. Forward strides in railroads, automobiles, and the steamboat, along with the invention of the telephone, the telegraph, and other modernizations, served to increase convenience and decrease arduous labor. Collectively, these innovations propelled the country into the next century while offering many entrepreneurs an opportunity to make their fortune. The financial benefits were great, but citizens across the nation began to realize the need for conservation. Two achievements of the time personify this movement, in my mind: the creation of the Palisades Interstate Park Commission and the construction of the Bear Mountain Bridge. The PIPC provided the means to embrace modernization but control the development of the Highlands in a thoughtful, productive manner. The construction of the Bear Mountain Bridge recognized the need for the valley to move forward, providing the access necessary for progress, not only for the betterment of the citizens of this localized area but also for the growth of the entire state, ultimately benefiting the entire country.

Early in the nineteenth century, admiration for nature began to develop, rising almost to the level of worship. Artistic movements such as the Hudson River School of Art and the Knickerbocker Group writers took hold in the Hudson Valley and boosted a growing appreciation of the natural surroundings. Both movements stimulated reverence for the ethereal beauty of the countryside as well as the meditative serenity available there.

Much of my research comes from firsthand sources: invaluable personal and business correspondence and original documents made available to me by the Orange County Historical Society, as well as other local Hudson Valley stewards of history; family information and photos provided by the families of Frederick Tench and William Welch; historical papers available from the Rare Book & Manuscript Department of the Butler Library, a part of Columbia University in New York City; memorabilia held at the historical archives of the Palisades Interstate Park Commission; pertinent data, both historical and current, provided by the New York State Bridge Authority (NYSBA); and so many other wonderful and generous sources that provided critical, factual information.

There were also areas of information that were more difficult to locate. Some of the biographical data on those individuals important to the story were available only through third-party sources—for example, biographical studies written by others that carry with them the opinions and impressions of their authors. In the late nineteenth century, during the "Gilded Age," it was in fashion to present images of those subjects of biography as more altruistic than reality might judge. As an example, Edward H. Harriman was a very controversial character of the times. After his death, several biographies were commissioned by the family, and while these books contained valuable information about him, it is understood that they might portray a biased image of the man. To counter this imbalance, I consulted writings by contemporaries with different perspectives, as well as newspaper articles of the day, in an effort to portray as accurate an image as possible.

I have attempted to present the main characters in this history as real human beings: neither saints nor sinners. In examining their impact on history and our world today, I hoped to give the reader a glimpse of who they were as people, rather than simply the effect they had on the events of this story. My fascination with history is owing in large part to my fascination with people and what makes them tick.

As we travel through this wonderful saga, we will watch Frederick Tench, Roland Harriman, George W. Perkins Sr. and Jr., William A. Welch, Elizabeth Vermilye, Cecilia Gaines Holland, Howard Carter Baird, and others who played significant roles in this saga fight to shape a world they believed in. They confronted political obstacles, resolved personal conflicts, raised funding, dealt with public opinion, and ultimately faced the challenges of the actual bridge construction.

I found the story of the Palisades Interstate Park Commission fascinating, and the people involved in its creation no less compelling. The coming together of so many dedicated residents, from a variety of backgrounds, pulling together to achieve the same goal, captivated me and, in my mind, served as the foundation story of the area and the bridge. Many of those persons who fought to advance the PIPC and its mission were residents of the valley early on, and, ultimately, it was either the same men and women or the members of their families who later advocated for the Bear Mountain Bridge. The chapters that follow will detail who those people were and the ingenuity and persistence they expended to eventually emerge victorious.

Steel and Grit chronicles how the residents of the valley joined together at a time when overindustrialization threatened to change their environment and how they created a permanent protector in the PIPC, thus ensuring the continuity of the lifestyle they cherished. Their energy and dedication ultimately culminated in the building of one of the most elegant and, at the same time, most technologically advanced bridges in the country—a bridge that has remained a soaring symbol of its era, personifying those efforts not only for themselves and their contemporaries but also for future generations. Few regions can boast the same amount of acreage preserved solely to allow residents to experience nature's serenity and untouched magnificence.

I invite you to come along with me on this journey through time. With any luck, you too will experience the same sense of enchantment I felt the day I passed under the Bear Mountain Bridge, slowly making my way through the timeless beauty of the Hudson River and its surrounding valley.

2. West Point marching band leading a parade of cars across the Bear Mountain Bridge during opening ceremonies. Courtesy of Frank Goderre Archives.

1

You Are Cordially Invited to the Opening

November 26, 1924

The parade of cars carrying the many guests invited to the opening-day festivities for the Bear Mountain Bridge reached the eastern-approach roadway of the newly completed span just as the fifty-piece West Point Band broke into "The Star-Spangled Banner." The automobiles pulled to a stop, and the occupants began to disembark and walk about, taking in the spectacular views of the Hudson River laid out before them. Thirty-five minutes earlier, they had assembled at the Peekskill train station, and then, following directives, they formed a procession of motorcars and made their way toward the bridge. The long cavalcade advanced, exiting the railroad station at South Street and moving slowly along Division Street to Highland Avenue, around the State Camp, up the new highway, past the toll gate, and finally on toward the bridge.[1] Each car was decked out in red, white, and blue streamers fluttering in the breeze.

Anticipation permeated the air. After twenty months of construction, the Bear Mountain Bridge—the first Hudson River crossing south of Albany—was ready to open to the public! At 1,632 feet, it was the longest suspension bridge in the world—a title it would hold for nineteen months before the Benjamin Franklin Bridge in Philadelphia took the title for itself.[2] Those guests invited to the ceremonies were, no doubt, feeling very smug as they presented the card, issued as part of their invitation, that gave them access to the bridge and roadway approaches a day before they opened to the general public. Special travel arrangements had been made for the 110 guests arriving from New York City; two coaches were added to train number 1 on the Erie Line, leaving Grand Central Terminal at 10:00 a.m. that morning and arriving at the Peekskill station at 11:10 a.m. Charles J. Donohue and his partner, H. Field Horne, Peekskill's local Studebaker agents, had been entrusted with providing enough Studebakers to transport those guests arriving by train and any residents who did not own a car. (Some guests drove from the train station to the ceremony in their vehicles, and if they had room for additional passengers, they welcomed them.) All told, Horne and Donohue provided thirty-one Studebakers.

Once the anthem finished playing, the guests were asked to please return to their vehicles and line up four abreast. At exactly 12:00, the band started to march. An assortment of festive tunes was played as they crossed the bridge with the parade of cars following closely behind. As the West Point musicians reached the west side of the span, everyone came to a halt. The cars positioned on the bridge measured forty lengths long, a total of 160 vehicles.[3] All passengers then disembarked once again and gathered around the west-end tollhouse. Two large American flags draped the ends of the bridge, and hundreds of smaller flags representing different nations around the world were strung end to end along each side of the bridge, further producing a festive atmosphere.

All eyes were directed to a small platform near the western tollhouse where E. Roland Harriman, president of the Bear Mountain Hudson River (BMHR) Bridge Co., was standing ready to begin the ceremonies. Roland joined Frederick Tench, the bridge contractor, to champion the project after Tench solicited his assistance back in 1921. As a principal in his brother's financial firm, W. A. Harriman and Co., Roland was able to organize a group of investors that evolved into the Bear Mountain Hudson River Bridge Co., the entity that ultimately funded and managed the project. On the platform, alongside Roland, stood his mother, Mary A. Harriman, widow of the late railroad financier E. H. Harriman. Also present was Wilson Fitch Smith, the chief engineer who had overseen the construction on behalf of the company. Harriman opened the festivities with a speech, giving the crowd a well-detailed background of the bridge construction and progress, which, he made clear, came in fits and starts. Then Mrs. Harriman, carrying a beautiful bouquet of red roses, proceeded to an area draped with two American flags and ceremoniously pulled a cord revealing to the crowd a bronze plaque mounted on the bridge.

The inscription read:

BEAR MOUNTAIN BRIDGE

THE FIRST HIGHWAY BRIDGE TO SPAN THE HUDSON RIVER SOUTH OF ALBANY
BEGUN MARCH 24, 1923 – OPENED NOV. 27, 1924

TO ALL WHO
WITH THOUGHT LABOR AND LOYALTY HAVE
CONTRIBUTED TO THE CONSTRUCTION OF
THIS BRIDGE AND HIGHWAY
THIS TABLET IS INSCRIBED

TOTAL LENGTH OF BRIDGE 2257 FT.	LENGTH OF SUSPENSION SPAN – 1632 FT.
HEIGHT OF TOWERS 355 FT.	CLEAR HEIGHT ABOVE RIVER – 153 FT.
DIAMETER OF CABLES 18 INS.	NUMBER OF WIRES IN EACH CABLE – 7252[4]

Smith stepped up to the podium and introduced himself to the crowd. He praised the artisans and workers involved in the project: "The ceremonies you have just witnessed here were but simple acts," he declared. "But they are important acts in the life of the Empire State. They signal the opening of an important course of traffic and the removal of the barrier which has blocked the southern portion of the State."[5]

Amid the high-pitched chatter that immediately followed the speeches, the West Point Band again took up their instruments and began to play. At the same time, newspaper reporters and photographers at the scene snapped photos of the Harrimans and other assembled dignitaries. Motion-picture photographers were strolling about, capturing the event from start to finish for posterity.[6]

During the previous two or three years, local papers, such as the *Peekskill (NY) Highland Democrat* and the *Middletown (NY) Daily Herald*, as well as the prestigious *New York Times*, had been filled with nothing but talk of the bridge, providing much fodder for speculation among residents. Initial stories reported on little more than the issuance of state and federal approvals for the bridge, but from the first suggestion that a bridge might be built across the Hudson at the chosen site, gossip began to circulate. Conjectures about the bridge's specifications regularly appeared along with assumptions about the cost and the source of funding. Much of the information was preliminary and often varied according to which publication was reporting. There were unending discussions on what the impact of the bridge would be on the immediate region and the entire state. It was reported that for the first time, citizens would be able to cross the river without waiting in line for ferries—which, of late, were having difficulty keeping up with the volume of automobiles and during the winter were not available when the Hudson was frozen. Access to the parks on the western shore of the Hudson would soon be easily available to residents of New York City and the surrounding towns and hamlets. The impact on commerce was expected to be great as well: now goods arriving at the rail terminals in New Jersey would be immediately loaded onto trucks and proceed on the journey to New York City without the need for loading and unloading onto and off barges to cross the New York Harbor.

Once the design was released, there was no shortage of opinions as to the beauty of the monument (or lack thereof) and whether it was worthy of its place on the river.[7] After construction began on the bridge in March 1923, the stories exploded, offering new reports to readers as each phase of the project was completed.

The project had been plagued by delays during the first sixteen months of construction, but a remarkable turnaround occurred in the summer of 1924 when job reports noted that a new, accelerated pace had taken hold, and one by one each requirement fell into place. A hint that completion was imminent came in August 1924 when Roland Harriman wrote to council members of the Peekskill community regarding road repair. In a letter to Cornelius A. Pugsley, dated August 11, 1924, Harriman expressed his concern that local roads should be inspected and repairs made so they would be ready to handle the increase in traffic the new bridge was expected to bring, thereby preventing any "neck of the bottle" conditions.[8] Pugsley was a lifelong resident of Peekskill, an active member of the local government, as well as an investor in the bridge, and an officer of the BMHR Bridge Co.

By the end of October, it was clear that the bridge would be ready within the next month. Roland wrote to the governor of New York, Alfred E. Smith, and informed him that the bridge would open for traffic on Thanksgiving Day, November 27, 1924, and the Bear Mountain Hudson River Bridge Co. was planning to hold an event the day before to celebrate the occasion. He asked the governor to "officiate at these ceremonies and to be our guest of honor at a luncheon to be served at the Bear Mountain Inn directly afterward."[9] Unfortunately, the Honorable Alfred E. Smith was already otherwise engaged and therefore had to decline the invitation. Roland continued to manage the list of invitees, often writing personal notes to close friends and dignitaries, asking several of them to speak a few words at the luncheon. In addition to Governor Smith, dignitaries solicited by Roland included Herbert Hoover, secretary of the interior; John W. Weeks, secretary of war; General F. W. Sladen, major general and superintendent of the United States Military Academy at West Point; and Benjamin B. Odell, former governor of New York, to name a few.[10] The resultant guest list numbered in the hundreds. Plans moved forward with anticipation and excitement.

The hundreds of news articles that had appeared over the years about the bridge, to say nothing of the rumor mill, fed the mounting excitement and anticipation of the locals. After all the talk, the bridge was now a reality. The following morning, at 7:00 a.m., Thanksgiving Day, the bridge would finally be open to public traffic. But today those special few invited to the bridge's inauguration would gather to celebrate!

The ceremonies at the western tollhouse concluded. As everyone returned to their cars, the parade continued with the band leading the way. The vehicles, now moving slowly westward in a single line, arrived at the traffic circle on the western shore. The band stepped to one side as the automobiles proceeded to the Bear Mountain Inn where a lavish luncheon was to be served.[11]

The menu was as follows:

> Fruit Cocktail
> Pickles Celery Olives
> Bisque of Tomato
> Roast Fresh Maryland Turkey
> Cranberry Sauce
> Fresh Brussels Sprouts
> Mashed Turnips
> Candied Sweet Potatoes
> Sweet Cider
>
> Fresh Pumpkin or Apple Pie with Cheese
> Cigars Coffee Cigarettes[12]

The inn was decorated for the occasion with spruce and evergreen sprays interspersed with twinkling lights, apropos of the season. Once all were seated and the meal was under way, Harriman rose to the microphone with a few jovial remarks. He then announced the names of thirty men who had participated in the bridge's success. As each man's name was called, he rose and accepted his acclaim. Those individuals included:

- Frederick Tench, partner in Terry & Tench Co., the contractor that ultimately built the bridge. Tench had initially conceived the project and approached the Harrimans for funding, then continued to administer the project and secure the state and federal approvals, and was awarded the contract to do the construction in March 1923.
- John V. W. Reynders, H. C. McCann Co., called in to supervise the entire project on behalf of Terry & Tench and Globe Insurance, in response to problems occurring in early 1924. He arrived at about the same time W. F. Carey and Company, Inc., was hired to address delays on the eastern-approach highway. A graduate of Rensselaer Polytechnic Institute, Reynders was active in the steel industry, managed Pennsylvania Steel Company, and built bridges in the United States and abroad. He was a member of the

American Institute of Mining and Metallurgical Engineers and was president of that body from 1925 to 1926.
- George W. Perkins Jr., current commissioner of the Palisades Interstate Park Commission, and friend of the bridge. Perkins invested in the bridge and served as cosignatory for Terry & Tench on their performance bond. The Perkins family was well known and active in the Palisades Interstate Park Commission, in the Hudson Valley, and throughout the country.
- Major William A. Welch, general manager and chief engineer for the Palisades Interstate Park Commission, was a consulting engineer for the bridge-approach highway. Welch served as an engineer in the US Army Corps of Engineers early in his career and again during World War I. He was very active in the movement to develop national parks and was known as the "father of national parks."
- Holden D. Robinson was the lead engineer on the suspension work for the Bear Mountain Bridge. He also worked on the Williamsburg and the Manhattan Bridges, as well as several bridges in Canada, Europe, and South America. Robinson was also an inventor and pioneered machines used in cable wrapping and squeezing, which helped to reduce the time required to build suspension bridges.
- Howard Carter Baird designed the Bear Mountain Bridge and served as a consulting engineer during the construction.[13] Baird had worked on many bridges throughout the eastern United States and New York metropolitan area for the Phoenix Bridge Company. He was a partner in Boller and Hodge, an engineering firm in New York City, and eventually took over the firm completely. He was a member of the United Kingdom's Institution of Civil Engineers, and the American Society of Civil Engineers.

As the announcement of the names concluded, the audience appeared to be confused. They realized that two of the most notable men involved in the bridge construction did not rise. Frederick Tench, the contractor whose initial idea and energy powered the project from conception through completion, and Howard Carter Baird, the engineer who designed the bridge and tailored it to its location, were not in attendance. Their conspicuous, unexpected absence posed a quandary for Roland and the others. After all the many months of hard work and determination, for some reason, these men were not on hand to accept acclaim from the many who had gathered to celebrate with them that day. But why?

As it turned out, the two men had been traveling together in Tench's automobile as they participated in the festivities. While crossing the bridge on their way to the banquet, Tench's vehicle broke down, resulting in the bridge's first traffic jam.[14] It was somewhat ironic and very unfortunate that the two men were cheated of sharing in the acclamations of their peers on this day owing to an unexpected incident occurring on the very bridge they were responsible for creating.

The dinner concluded, and Harriman offered some final remarks and then introduced former governor of New York Benjamin B. Odell, of Newburgh, who said a few words. Former chief justice of the New York Supreme Court Morgan J. O'Brien followed Odell, along with J. DuPratt White, of Nyack, chairman of the Palisades Interstate Park Commission and founding partner of Chase & White, a New York law firm, each taking their place at the podium and speaking to everyone in attendance. It was 2:45 p.m. when the ceremonies concluded, leaving just enough time for those people returning to Grand Central Terminal in New York City to catch the 3:00 p.m. train from the Bear Mountain train station.[15]

At 7:00 a.m. the following day, with toll takers waiting at their posts in anticipation, the bridge opened to the public. The toll for vehicles exceeding one hundred inches in length was one dollar, plus an additional fifteen cents for each passenger (ten cents for children). Vehicles less than one hundred inches were charged eighty cents, with the same passenger fees. Traffic did not start to get heavy until around noon. By 1:30 p.m. it was estimated that about fifteen hundred cars had passed over the bridge; by 4:00 p.m. the number had risen to five thousand. It was reported that at least a dozen police officers had been enlisted to regulate traffic throughout the day.[16]

The opening easily earned the applause of the area press. Indeed, the opening of the Bear Mountain Bridge was reported in multiple newspapers, not only the ones close to home but many across the country and all over the world. A letter to Roland, dated December 11, 1924, from Irving Rossi, a W. A. Harriman and Co. associate working in the company's Berlin, Germany, office, provided insight into how far the news of the bridge's opening had traveled. "I beg leave to enclose a picture of an American bridge. This picture was taken from a German newspaper," wrote Rossi. "I believe congratulations are in order. . . . [T]he newspapers I have received advise me it was opened on Thanksgiving Day." His clipping was from a Berlin newspaper that had praised the bridge and those individuals responsible for building it.[17] Among the many other congratulatory notes received by Harriman and the Bear Mountain Hudson River Bridge Co. were several from invitees unable to attend the unveiling ceremony, as well as many who had attended but wanted to express their thanks for being invited to share in the day.[18]

The completion of the bridge, and within the planned time frame to boot, was indeed an accomplishment, given the problems they had encountered early in the schedule. The eastern-approach road to the Bear Mountain Bridge was an immense challenge that almost undermined the entire project. This part of the construction may well have been the proverbial "straw that broke the camel's back" for Terry & Tench Co., Inc.—the impetus that precipitated the calling of the performance bond—an unfortunate result of financial problems the company endured. The road was cut out of the southern face of Anthony's Nose on the eastern shore at an elevation of more than 400 feet above the river. Almost the entire roadway required excavation and fill. Because of the solid granite terrain, 70 percent of the material had to be drilled and blasted to remove it.[19] In addition to the difficulties of excavation, there were issues relating to the incongruent elevations encountered. While an elevation of 200 to 300 feet was maintained as a standard along most of the roadway, the road climbed to more than 400 feet within a mile of the final approach to the bridge and then dropped abruptly to somewhere around 178 feet. This dramatic drop resulted in retaining walls from 15 to 40 feet, adding to the complexity of the project and presenting additional difficulties. At one time as many as sixty air drills operated by a battery of twelve air compressors were used—incurring unanticipated costs for added time, personnel, and equipment.[20]

Challenging though it was, the eastern-approach road turned out to be one of the most scenic components of the project. The three-mile drive winding along the southeastern side of Anthony's Nose, high above the Hudson River, affords spectacular views of the Hudson. It's been reported that one who drives along this stretch of highway and continues over the bridge, north via Storm King Highway toward West Point, covers one of the most magnificent scenic routes in the eastern part of the United States.[21]

Indeed, this part of America has been compared to the most celebrated landscapes in Europe. In the December issue of the *Outlook* magazine, based in New York City, a writer remarked, "If the Hudson River were in Europe, Americans would be crossing the ocean to visit it. The scenery of the Hudson is equal to anything in the English or Scottish lake districts. It is comparable to that of the Rhine. The opening of the bridge across the Hudson just below West Point, therefore, is an event of more than local or even sectional concern."[22]

2
The Valley

The story of the Bear Mountain Bridge cannot be properly told without first telling the history of the surrounding valley and the people who lived there and worked to preserve the environment they loved. The centerpiece of the valley is the Hudson Highlands, the mountainous area flanking the Hudson River, encompassing Putnam County on the east shore of the river and continuing south to include Westchester County; on the west shore it contains Orange County and continues south to include Rockland County. The valley was formed during a bygone ice age when a precursor to the Hudson River entered a fjord and cut through various rock formations.

The Palisades, a continuous track of cliffs, runs from Jersey City, New Jersey, northward along the western shore of the river to Nyack, New York, through the Hudson Valley, for a total distance of approximately 20 miles. Like a line of erect, vertical stakes or disciplined soldiers, they stand at attention majestically straight and tall, reaching upward for 300 feet at their lowest summit, gradually rising to a maximum height of 540 feet at their northernmost point. The name Palisades comes from the Latin word *Palus*, or "stake." The first map of the New World, drawn by European mapmaker Gerardus Mercator, in the early 1500s, labeled them "Anorumbega," a shortening of the term Florentine explorer Giovanni da Verrazzano used to describe them. Verrazzano was commissioned by the king of France to find a passage to the Far East, and while doing so, he would chart the eastern coast of what was to become the United States. He called the New York Bay area "La Terre de L'Anormee Berge," which translates to "Country of the Grand Scarp."[1] In 1524, Verrazzano crossed the Atlantic, entered lower New York Bay, and sailed up the North River (later renamed the Hudson River), searching for the elusive passage to the Far East. As he passed the Palisades, he noted their splendor in his journal. It was Verrazzano's notes that gave Mercator the information he needed to produce his map, thus saving Verrazzano's discoveries for posterity.[2] For whatever reason, Verrazzano and his crew did not stay long in the bay that prompted his deep admiration. Perhaps they realized quickly that there was no access to the East through these waterways.

Many other explorers followed, sailing north along the river on behalf of various European nations, all searching for the elusive magical passage to the Far East, but most did not document their journeys and their findings were lost to future generations. In 1609, however, an Englishman by the name of Henry Hudson, sailing a Dutch vessel christened the *Half Moon* and manned by a Dutch crew, happened into the same river valley. He found it rich in resources and inhabited by Indigenous people. Both Hudson and one of his sailors kept detailed journals, and their entries have provided historically invaluable accounts of the journey, including interactions between the crew and the numerous tribes they encountered, some of whom proved to be amiable, while others were hostile and often violent.[3]

None of the explorers who set their course for Hudson Bay ever found that magical passage to the Far East, but many recognized the richness of the land they stumbled upon, especially the island that would later be known as Manhattan. With favorable reports winding their way around Europe, explorers continued to arrive in the region, and, eventually, settlers venturing from various European countries came to establish homesteads and settlements as they embarked on the adventure of starting a new life in a new world. In 1624 the Dutch, following Hudson's lead, settled the island of Manhattan as a part of "New Netherlands." The locals, whom Hudson's sailors had labeled "savages," or *wilden*, were wary of the trespassers. The tribes were numerous and diverse, and they received the intruders with varying degrees of cooperation. For their part, the explorers regarded the *wilden* with suspicion and distrust; the *wilden* responded in kind.

The Indigenous people known as the Lenape occupied much of what is now New York, New Jersey, Delaware, and Pennsylvania and referred to what would eventually be named the Hudson River as "Muhheakunnuk," or "great waters constantly in motion."[4] They were divided into many autonomous groups all related by a complex system of kinship that traced their roots back to common ancestors. The Europeans called one group the "Delaware" because they had immigrated to New York Harbor from the area around the Delaware River. The Lenape people, known as the Munsee because of the language they spoke, lived in the area that is now New York City. The Raritan tribe inhabited what is present-day Staten Island and central New Jersey, while the Tappan occupied the Palisades and the Haverstraw were at the bay that now bears their name. The Nyack settled on the eastern shore of the Narrows, with ties to the Hackensack, and the Canarsee lived in what is now Brooklyn. Many of these names are part of our everyday lives today. We've used them in labeling our streets, towns, and neighborhoods. Along the shores of the Muhheakunnuk were the Algonquin and the Mohawks, who were part of the Iroquois nation. The Algonquin and Iroquois were traditionally competitors and enemies and were known to be both fierce and violent fighters when provoked. Hudson and his crew had unknowingly ventured into a very hostile atmosphere.

The Dutch maintained dominance in the area through the early seventeenth century, but in September 1664, the British successfully invaded, taking possession of the settlements ruled by Peter Stuyvesant, the director general of New Netherlands. Over the next century, the British maintained a continuous presence along the coastline from Virginia to Maine and regularly fought battles with tribes of Indigenous people who remained in those areas as well as with the French, who had a stronghold in the northern territory close to where the Hudson River met the St. Lawrence River.

Bolstered by a steady supply of new settlers, sophisticated weaponry, and sheer persistence, English forces eventually overwhelmed the native groups and solidified their rule along the lower Hudson and in the Province of New York (named for the Duke of York, brother of England's King Charles II in 1664). This British colony roughly included present-day New York State and Vermont.[5] England ruled their colonies with an iron hand and maintained order and loyalty among their colonists. However, by the late 1700s the English colonists rebelled and won their freedom. The War of Independence gave birth to the United States of America. The decades that followed were tumultuous, with the War of 1812 following closely after the American Revolutionary War. Through continued exploration, the new nation grew, with the core thirteen colonies maintaining their place as the heart of the nation. Settlements expanded from this core area, and subsequent evolution and progress developed into surrounding territories. The valley along the Hudson River, located only miles north of Manhattan, was soon populated with settlers. The river was recognized not only for its beauty but also for the convenience it afforded as a means of travel and commerce. A copy of a book written in 1775 and published in London, titled *American Husbandry: An Account of the Soil, Climate, Production, and Agriculture of the British Colonies in North America and the West Indies*, authored by "an American," was discovered to contain a description of the "river Hudson" written with poetic flourish:

> The river Hudson which is navigable to Albany, and of such a breadth and depth as to carry large sloops, which its branches on both sides, intersect the whole country, and render it both pleasant and convenient. The banks of this great river have a prodigious variety; in some places there are gently swelling hills, covered with plantations and farms; in others, towering mountains spread over with thick forests; here you have nothing but abrupt rocks of vast magnitude, which seem shivered in two to let the river pass the immense clefts; there you see cultivated vales, bounded by hanging forests, and the distant view completed by the Blue Mountains raising their heads above the clouds. In the midst of this variety of scenery of such grand and expressing character, the river Hudson flows, equal in many places to the Thames at London, and in some much broader. The shores of the American rivers are too often a line of swamps and marshes; that of Hudson is not without them, but in general, it passes through a fine, high, dry, and bold country, which is equally beautiful and wholesome.[6]

In the early and mid-1800s, a group of landscape artists discovered the serenity of the Hudson Valley and identified it as a place of aesthetic inspiration. Thomas Cole, Jasper Francis Cropsey, George Inness, Frederic Church, and Albert Bierstadt memorialized the scenic treasures of the valley through their paintings. An awareness of the serene magnificence of the Hudson River and its surrounding valley—and the value in preserving its magnificence—slowly began to take root within the region. Eventually, the impact of these artists on the art world was so great that a new label was developed to describe their unique style: the Hudson River School.[7] At about the same time, a faction of writers—regular contributors to the *Knickerbocker* magazine—known as the Knickerbocker Group also served to promote the beauty of the valley and the importance of conservation. The *Knickerbocker* magazine, founded in 1833, was one of the first publications to provide a platform for a Romantic interpretation of the American wilderness among writers in New England and New York. With contributions from such prominent writers as Washington Irving, William Cullen Bryant, Henry Wadsworth Longfellow, Oliver Wendell Holmes, Thomas Cole, Francis Parkman Jr., and others, it is easily considered one of the earliest proenvironment magazines in the United States.[8]

New York City grew to be a major port, and as the population swelled, the need for food and building materials multiplied. Many of the products in demand were produced in the Hudson Valley, a short boat trip away. Steamboats became the main shippers of freight, and the waterways were constantly congested with steam tugs and scows pulling barges to and from the terminals. The valley produced coal, which was the main fuel used at the time, as well as Rosendale cement and bricks made in local brickyards; all of these items were shipped to New York City and used in building trades. Ulster County bluestone, shipped from Ulster County to the city for use as sidewalks, was in demand. Ice blocks cut from the river in winter were sent to the city to be used for food preservation. Grain as food products along with hay for horses were grown in the Midwest and then harvested and sent through the Erie Canal to the Hudson River and on to New York City via barges.[9]

The Hudson River waterfront was one of the busiest working ports. With the opening of the Erie Canal in 1825, the traffic only increased. The river yielded a plentiful catch of a variety of fish species and seafood. Oyster barges were a common sight on the water, and the city streets were filled with carts selling plentiful catches to hungry customers. In addition, many of the world's most celebrated ships came and went from piers in Chelsea and Hell's Kitchen, with soldiers, celebrities, immigrants, and high society traveling abroad. The *Carpathia*, which rescued survivors of the *Titanic* in 1912, docked at Pier 54 to deliver them home. The British steamship the *Lusitania* departed on its last voyage from a pier in New York City, before it was torpedoed by a German submarine in 1915.

The Hudson Valley's initial economic boom involved real estate, not industry. Those individuals who could afford to do so bought up available lots in the Highlands and built vacation homes, taking full advantage of the awe-inspiring views of the Palisades and the enhanced status they provided. The Palisades Mountain House, a resort hotel located on the southeastern slope of an elevated plateau of the Palisades, was one of the most sought-after resorts in the area and represented the epitome of luxury vacationing for New York society in the early 1800s.

At the same time, fishing villages popped up along the base of the Palisades, housing local fishermen who made their living from the plentiful catches of shad, sturgeon, cod, and other species that inhabited the river. The Hudson River is an estuary—a unique marine environment where freshwater and saltwater mix (brackish water). Depending on the tides and the seasons, the two combine in varying proportions for up to sixty miles north of New York Harbor. Because many saltwater species of fish require fresh or brackish waters in which to lay their eggs, the Hudson provides an ideal spawning ground.[10]

During the late eighteenth century and continuing into the early nineteenth century, many new technological advancements appeared, shaking up the varied industries as the first Industrial Revolution took hold. Strides were made in textile manufacturing, machine tools, and metallurgy, and steam as a source of power was introduced to the labor market.

Between 1830 and 1860, the United States experienced a railway building boom. Although the use of steam as a source of power had been discovered in the early eighteenth century, it was not refined and truly utilized until decades later. The first steam locomotive was introduced in 1804, and the first steamboat appeared in 1807 when Robert Fulton introduced the *Clermont*. These breakthroughs in transportation effectively shortened the time necessary to travel long distances, and they began to appear across the country.

In 1840 a plan arose to construct a railroad along the eastern shore of the Hudson River. Initially, it was met with both political and physical opposition. The terrain presented several engineering challenges that were certain to increase the cost of construction significantly: mountains had to be tunneled, marshes had to be stabilized, and substantial amounts of rock fill had to be brought in as a base to allow the track to be laid. At the same time, political opposition led by the backers of an existing competing line, the New York and Harlem Railroad, was brewing in Albany, and there was strong opposition from the steamboat operators as well. The mid-1800s was considered by many historians to be the "golden age of steamboating" in the Mississippi and Ohio Rivers, and the conditions in the Hudson River were no different. Steamboats enjoyed what amounted to a transportation monopoly. Frequent runs by steamboats transporting passengers and freight were evident up and down the river since at the time they were the most efficient means of commerce. Indeed, one could scarcely observe the river without a steamboat in view.[11]

Nevertheless, in 1842 a group of undeterred investors presented a proposal for a railroad charter to the New York Legislature. Steamboat interests were successful in stopping them on their first try, but three years later the investors regrouped with a new plan: this time they employed John Bloomfield Jervis, America's leading consulting engineer of the time (antebellum era, 1820–60), to take the lead on the project. Jervis did a survey of the route and a market analysis of the venture, and at the same time, he became the public face of the project. He was a pioneer in the railroad and canal industries and was recognized throughout the country. He designed five of America's first railroads and supervised their construction, served as chief engineer on three major canal projects, and designed and built the Croton Aqueduct that provided the fresh water supply for the residents of New York City from 1842 to 1865. Jervis was also an inventor and was responsible for designing several of the early steam locomotives. Many of his original designs are preserved today at the Smithsonian Institution as well as at the Library of Congress in Washington, DC.[12]

Jervis, himself, took the revised proposal before the New York Legislature, singlehandedly facing off against the opposition. It took him three months of presenting arguments before he finally secured the approval.

After getting the charter approved, he and his backers proceeded to issue stock, only to find that while many supported their cause, few were willing to commit financially to the project. Finally, following a yearlong publicity campaign, they were able to raise the $3 million necessary to bring the dream to reality.[13] Work began in 1847, only to be met with yet another obstacle—and this one seemed insurmountable.

At Anthony's Nose (a mountain on the river's east bank and the future site of the Bear Mountain Bridge), the only way to maintain a straightaway was to tunnel through the mountain. But drilling through the granite was believed to be impossible. One contractor after another took on the challenge only to ultimately admit defeat and leave the job. Finally, a contract was awarded to H. D. Ward and Company. After seventeen months of intense drilling, the tunnel was completed.[14]

It was late in 1849 when passenger service on the Hudson River Railroad finally commenced. Passenger traffic increased because the train ran regardless of the weather—something the steamboat could not offer. Additionally, the train ride from New York City to Albany was only five hours—half the time it took a steamboat. However, the freight business was much more difficult to sway. In response to the railroad's efforts to attract the freight business, the steamboats cut their prices and engaged in a rate war. This rate war, coupled with the established habits of freight shippers, may account, at least in part, for the fact that the railroad could not break into the freight business for several years. As a result of this and various management issues, the railroad did not show a profit for almost fifteen years. This fact, however, doesn't tell the whole story. Five years after the Hudson River Railroad began service, a man named Sam Sloane assumed the position of president of the railroad. He brought with him strong management skills and a prudent frugality that allowed him to slowly turn the company's balance sheet from negative to positive. During his time at the helm, the railroad's stock price rose from $17 per share to $140 per share.[15] Ultimately, by 1869 the Hudson River Railroad was considered a successful venture as it merged with the New York Central, where it remains today.

Still, the eventual success of the railroad did little to impress valley residents who were unhappy with it from the start and resented the railroad for allegedly desecrating the valley landscape. Their sentiments reflected the emerging movement toward conservation and concern for the environment.

As the Hudson River Railroad's saga showed, the increase in railroads brought with it a need for bridges to carry the rails across the rivers. By the mid-1800s the span from Bear Mountain to Anthony's Nose had been surveyed and investigated three times for a river crossing because of its ideal topography. But it was not until the Bear Mountain Bridge was built, in 1924, that success was achieved.[16]

Before the Bear Mountain Bridge, the bridge project that had come closest to completion was an 1868 proposal for a railroad crossing initiated by an association between the Erie Railroad and the New England Railroad Companies. The intention was to extend their railroads to connect Turners Station (later known as Harriman Station), located west of Newburgh Junction on the Erie Railway in New York, across the river to the town of Derby in Connecticut. The association brought its petition to the New York Legislature and secured the necessary approvals to extend the tracks of the railroads involved, along with the rights to build the required bridge. General Edward W. Serrell was named as engineer in chief for the project and was entrusted with the selection of the location as well as the preparation of the plans. A Union veteran, General Serrell was well known for his engineering feats during the Civil War and his many useful inventions such as armor plates, impromptu gun carriages, and other apparatuses that contributed to the war effort. He worked as a consulting engineer on many railroad and canal projects. Among them were the Erie Railroad, the Northern Railroad of New Hampshire, the Central Railroad of New Jersey, and the Union Pacific (UP). He was also involved in constructing the suspension bridge across the Niagara River at Lewiston and the bridge at Saint John, New Brunswick, and he planned the bridge over the St. Lawrence River in Quebec. In 1879, he was called upon as a civil engineer to examine the plans and sections of the Brooklyn Bridge, then under construction, and to testify relative to the calculations on the assumed strength of the bridge before the Assembly Sub-committee on Commerce and Navigation. He was also listed as an honorary pallbearer at the funeral of Ulysses S. Grant.[17]

Serrell planned to work with Colonel M. O. Davidson, chief engineer of the New Haven and Derby Railway, who was to preside over the part of the Erie/New England project located in Connecticut.[18] On June 13, 1868, *Harper's Weekly* published a rendering of the planned bridge, giving the public a peek at what they could expect. It informed readers that per Serrell's recommendation, the bridge would cross the Hudson at Fort Montgomery on the west shore to Anthony's Nose on the east.

The project stalled for some reason—most likely a financial snafu—but was resurrected twenty-two years later, leading *Harper's* to publish another enthusiastic review of the plan on May 10, 1890, with a description that may have differed only very slightly from the 1868 details; in most ways, the project essentially seemed the same. The article confirmed that "the towers will have the appearance of granite . . . and double track railways of standard gauge will occupy the upper deck, the lower one will have a roadway and footpath for local traffic."[19] The bridge, which required as part of its design a tunnel through what was known as "Bull Hill" on the western extension of the rail line, was projected to be ready in two years. But despite all the planning, and the hard work and preparation by Serrell's team, the bridge was never built.[20]

The many attempts to build a railroad bridge at this location testify to the all-encompassing drive to expand the railroads across America and the importance of the railroad to the country's growth after the Civil War. A major effort was under way to cross the country with iron tracks and spikes. The surge of railroads in the nineteenth century had a strong impact on the everyday lives of Americans. The time necessary to travel across this vast country was reduced significantly. In effect, the country became smaller. Remote locations could now be accessed for settlement and commerce, and many of the innovations developed then would be available to those persons in otherwise isolated territories.[21]

The advances in transportation provided people with more convenience in travel and opportunities to relocate more easily than ever before, thus instigating a movement away from the cities and toward the more rural areas surrounding them. These rural areas were eventually dubbed the "suburbs." Nathaniel Parker Willis, a controversial magaziner of the nineteenth century, suggested that the villages along the Hudson River were indeed the "suburbs" of New York City. He noted, specifically, that steam had shortened the distance between those villages and New York City and that New Yorkers would find "that they could carry on their business in the city, and at the same time provide a cheaper living for their families in the country."[22]

By the middle of the nineteenth century, the Second Industrial Revolution was under way, and additional technology was initiated in the form of electricity, oil, and gas; uses for steam power were being further developed and utilized in many ways. Communication improved with the inventions of the telegraph and the telephone. Transportation moved ahead using steam to power railroads and steamboats, and further strides were established through the implementation of the automobile, and eventually airplane flight as a result of the efforts of the Wright brothers in North Carolina.

The years following the Civil War saw the United States experience an economic boom, and with the approach of the Second Industrial Revolution old industries expanded while many new industries were born. These newer innovations resulted in opportunities for investment and the potential for making money. Raw materials were needed, and entrepreneurial individuals lost no time locating and providing them to the manufacturing sector. Across the country, as well as in the Hudson Valley, the increased activity was initially thought of as a positive indication of promised prosperity. However, residents soon recognized the dangers that could result if entrepreneurs were allowed to continue exploiting natural resources without implementing plans for conserving or renewing them, or both.

In Hudson Valley, locals tried to prevent these enterprising outsiders from buying up land, but it was difficult. Passionate residents voiced their outrage as the local resources were harvested before their eyes. They petitioned their political representatives, and local newspapers carried reports of crusaders supporting the cause of conservation.

The federal government also recognized the need for uniform legislation to standardize protections for all American citizens. Federal agencies, specifically designed to protect natural resources for the benefit of future generations, began to appear. With the creation of these agencies came many new regulations. Between 1879 and 1910, numerous state and federal laws were passed as protection against overindustrialization and overdevelopment. The US Congress established the United States Fish Commission in 1871, providing, for the first time, a full study of the ecology and resources of the country's wild fish, shellfish, and marine life. The investigation was done for the protection of these species and to ensure their propagation; the commission was ultimately absorbed, in 1939, into the US Fish and Wildlife Service, an agency under the US Department of the Interior. The US Geological Survey was formed in 1872 to monitor, collect, and analyze scientific information on natural resource conditions; its mission was to identify critical problems and find solutions for them. In 1881 the federal Division of Forestry was created, and within a decade the United States had established a formal system for managing our national forests.[23]

While the government created formal bureaucracies, citizens joined together to create public interest groups. Between 1875 and 1905, those citizens joined clubs such as the American Forestry Association, the Sierra Club, the Appalachian Mountain Club, and the National Audubon Society, to name a few. As perspectives evolved, two distinct movements were born: one prioritized conserving renewable resources for quality of life (for example, forests, wildlife, water, and fisheries), and the other focused on preserving the wilderness for spiritual value.[24] Additionally, there was both a recognition of historical memory and the desire to preserve historical landmarks to enlighten and educate a new generation on the events that had occurred and had shaped Americans' current existence. Men like Washington Irving and William Thompson Howell rose as leaders in this movement. They worked to convince others that historic landmarks should be preserved and saved from falling into ruin. With the help of the concerned public, legislation was generated and passed, resulting in the purchase and restoration of several sites.

While the growth of railroads was perceived as a positive development for the country to improve transportation, there were issues of a negative impact. Hudson Valley residents felt that the beauty and serenity of the valley were being damaged, and many were against further expansion. The ease of transportation also allowed more entrepreneurs with ideas for making money to arrive in the valley. These industrialists typically brought ideas that involved depleting valley resources without any restoration plan. Urged on by their constituents, politicians formulated an assortment of ideas. Some suggested forming a commission or organization and arming it with the authority to seize property owned by the quarries or others who were engaged in defacing the cliffs. Still others offered that the solution was to form a park, owned and managed by the state, along with a grand boulevard to be located along the crest of the cliffs, thus offering residents a promenade with magnificent views of the river, as well as Manhattan and Brooklyn in the distance. The suggestions of a park and grand boulevard, of course, assumed the use of properties in need of protection and therefore would render those properties unavailable to anyone who wished to deplete their resources.[25] Articles appeared in many of the local New York and New Jersey newspapers, feeding the debate and supporting the need for a solution. And as the battle for conservation raged within the Hudson Valley, it erupted across the country.

In February 1894, an article appeared in the *Jersey City News* announcing that a movement establishing a private park on the Palisades was gaining momentum. The local landowners were concerned that the newly constructed Palisades Railroad, then completed as far as Fort Lee but expected to reach Englewood within the year, would bring a deluge of land boomers and auctioneers anxious to buy up available real estate and sell it off in lots. The park plan was formulated by local banker George S. Coe in cooperation with Judge William Walter Phelps, both landowners in the valley. The establishment of a park, they reasoned, would fight this outcome and preserve the rugged beauty of the area. The newspaper article reported that "an agreement is being circulated for signatures to form an unincorporated association of mutual benefit, the members agreeing to work together for the improvement of their lands within the territory specified and to pay an annual assessment of $1 per acre to maintain and improve the property." As of the date the article appeared, already one-quarter of the landowners had signed on to the agreement, introduced only a short time earlier. Expectations were that everyone within the designated area would cooperate. This early plan was being managed by William E. Bond of No. 2 Wall Street, New York, who was the confidential representative of Judge Phelps.[26] Although we do not have the outcome of this specific plan, it demonstrates the sentiment and determination of the valley residents at the time. It seems like a precursor to the formation of the Palisades Interstate Park, which did not come until years later, after much effort and diligent work were expended, resulting in the legislation of 1900.

In September 1894, New Jersey's state geologist, John Smock, declared that he was investigating the issue and would soon submit his report to Governor George Werts for review. Smock said that he favored the creation of a commission in New Jersey similar to the "Metropolitan Park Commission," which had been formed by the Massachusetts legislature. The commission would address issues like providing open space within cities, preserving attractive scenic features, and, most important, addressing the need for parks in New Jersey. Smock pointed out, "I have been watching the Palisades for several years. . . . While the destruction is not great at the present time, it is the beginning of what means mischief to the scenery of the Hudson River." Smock formed a committee that included Rev. Samuel B. Dodd of Hoboken along with Wendell P. Garrison, editor of the *Nation* and resident of Orange, and Colonel W. H. Wheeler of Hackensack to assist him in the investigation and compiling of the report. Smock also consulted with the New Jersey attorney general, John P. Stockton, who confirmed that the right of condemnation lies with the legislature. Stockton specified that should the supporters of a commission wish to grant that body the right of condemnation, they would need to solicit and be granted that power from the legislature. Adjutant General William B. Stryker, who had his summer house on the Palisades and lived there for a portion of the year, also commented that not only was the natural beauty of the Palisades being destroyed, but so were the "comfort and quietness of the citizens as well. In the mornings at six o'clock and the evenings at sunset, the cannonading underground goes on, and it can be heard for miles." General Stryker had been responsible, in great part, for the passage of a law in 1874 that made it a misdemeanor to place or paint on the Palisades anything considered an advertisement. Discussions now developed that suggested that this law be amended to cover the current issues, making it illegal to deface or destroy the cliffs. Some of General Stryker's neighbors were George B. Coe, banker; J. T. Lamb, maker of cathedral-stained glass; J. Cleveland Cody, New York broker; and William S. Opdyke, lawyer.[27]

Not everyone agreed that the horror of defacement should be a reason to end the destruction. As there always are, some residents held varying opinions on the matter, which were also reported in the local newspapers. *New York Tribune* reporters spoke to some who had no care if the cliffs were demolished. One such citizen remarked, "Why, of course, they should continue the work, and if I had my way they'd never stop till the hull [sic] darned mountain was down. There it stands doing nobody any earthly good, interfering with our communication with New York, and the only people who want the stone there are the scenery lovers." Those residents who objected to the blasting were often put down as "sentimentalists." Those individuals who supported the blasting argued that at least the stone was being used for practical purposes, such as building and paving. The conflicting opinions were the source of much discussion, during which another question arose: the determination of legal ownership. Deeds for many properties located on top of the Palisades showed ownership only to the edge of the cliffs, and properties located along the shore claimed the land from the low-tide mark up to the base of the Palisades, leaving a question as to who actually held ownership of the face of the rock. The question remained under dispute while the blasting continued.[28]

Additionally, many residents expressed concerns for safety owing to the large quantities of dynamite being used. The blasts were becoming greater and more powerful, and people were fearful that the destruction would spread to adjacent private properties.

One report told of an owner petitioning the grand jury in Hackensack, New Jersey, to stop the blasting because it was impacting her ability to secure summer renters for her property. Large stones were being hurled onto the property by blasts being done by the Carpenter Brothers quarry.[29] The Carpenter Brothers quarry was located approximately two miles north of Fort Lee on property owned by William O. Ellison in Englewood, New Jersey. While the contractors issued assurances that there was no danger to adjoining properties and occupants, people nearby were still concerned about the risks.³⁰ The *Perth Amboy (NJ) Evening News* reported a lawsuit heard before Justice Keogh in New Rochelle, New York. Mrs. Clinton S. Arnold, of Ossining, New York, brought suit against the Rockland Lake Trap Rock Company, commonly referred to as the Odell-Foss Quarry Company. Mrs. Arnold claimed that because of heavy blasts set by quarrymen directly across the Hudson from her house, the house was severely shaken, the walls were cracked, and other damage occurred. At the trial, she brought evidence supported by her husband, an architect and builder, and was supported by other residents of the village of Ossining. Expert testimony by scientific geologists was entered, claiming that the ledge of rock being blasted on the west side of the Hudson passed under the river and continued into the eastern shore, and because of that fact the heavy blasts set off at Hook Mountain (on the west) directly caused shaking houses and resultant damage in Ossining and other towns along the eastern riverfront. The suit alleged that all of Westchester County, plus other sections of the Hudson River valley, were being affected by the blasting. Several neighbors testified to various problems they had experienced because of the blasting. Some had ceilings fall, furniture damaged, and decorative items smashed beyond repair as a result. Dr. Irvine, the physician to Sing Sing prison inmates, told the court how the prison was also jarred by the recurring blasts from across the river.³¹

Late in 1894, as more articles appeared in the *Jersey City News*, residents of New Jersey continued to push for a solution to the problem. A special session of the state legislature convened in October, and Senator Henry D. Winton of Bergen County was present in Trenton to present his bill, one of the first pieces of legislation addressing this issue. However, the legislature adjourned immediately upon assembling, and while Senator Winton attempted to get his bill heard, he was unsuccessful. Stubbornly, he vowed not to give up; he would introduce the bill again at the January session. The bill outlined a proposal that Governor Werts appoint three commissioners who would investigate the matter and report on recommendations for protecting the natural scenery of the Palisades, possibly including the creation of a park. It appeared that the destruction had been occurring for many years, but the previous decade had seen a considerable increase in quarries destroying the cliffs. Senator Winton and Governor

Werts had traveled to the Palisades to see for themselves the amount of destruction being done. Governor Werts agreed with Senator Winton that something must be done and confirmed that he would approve Winton's bill. Another resident of the area, Mr. Fleming P. Albert, whose property was located on Pulpit Rock directly across the river from Yonkers, was also a vocal advocate for saving the Palisades. Albert had given the matter a lot of study and was knowledgeable about the history of the area. He was also in favor of a grand boulevard being constructed along the entire twelve miles, starting at Fort Lee and ending at the state line.[32]

Another supporter, Frederick Law Olmsted, well known as the father of landscape architecture in the United States, and the designer of Central Park in New York City and Prospect Park in Brooklyn, was quoted as saying: "There is one place on the Hudson River which ought to be held by the public.... I refer to the Palisades.... It will in time become a place of great magnificence.... The blasting should be stopped at once."[33]

In 1895, a milestone for scenic conservation was reached when a charter was drawn up to create the Trustees of Scenic and Historic Places and Objects in the State of New York. In 1901 the name of the organization was changed to the American Scenic and Historic Preservation Society. It was recognized as the first agency in the state to act as a trustee for scenic, historic, and scientific properties, adopting conservation as a firm practice. For the first time, historic landmarks were in the hands of those people who would conserve and maintain them.[34]

Theodore Roosevelt, who served as governor of New York from 1899 to 1900, and as the twenty-sixth president of the United States from 1901 to 1909, was one of the first officeholders to publicly champion conservation. His dedication to preserving the land is illustrated by his words: "To waste, to destroy our natural resources, to skin and exhaust the land instead of using it to increase its usefulness, will result in undermining in the days of our children, the very prosperity which we ought by right to hand down to them amplified and developed."[35]

Faced with the possible devastation of their environment, residents of Hudson Valley continued to resist. Average citizens instigated the movement, but they were soon joined by influential families who called the valley home. Among them were the Perkins family, the Harrimans, the

Rockefellers, and the Vanderbilts, who lent their support to the efforts. Other families included the Morgans, Partridges, Howells, Bigelows, Olmsteds, Osborns, and Stillmans; all were well known in the valley, and all played a big part in this movement. Many of these families had lived in the valley for generations, and they had the financial means as well as the relationships with those individuals who could garner the cooperation of local and state governments to accomplish things.

Many of the people listed above were captains of industry who reaped many benefits from the industrialization taking over America, prompting a concern about inconsistency. However, they may have recognized that while progress, in general, might prove beneficial for the residents of the Hudson Valley, the destruction of the Palisades was not progress, but rather the result of irresponsible practices. All the residents of the valley, rich or poor, reacted to the blatant recklessness of those persons who gave no thought to the replenishment of what they took and seemed oblivious to the travesty of dynamiting cliffs that had stood for centuries and were irreplaceable. The aim was to support progress, but control and temper it with responsible management.

The patriarch of the Harriman family was Edward Henry Harriman. Henry was a controversial figure—a railroad baron who loved the outdoors and wanted to preserve its natural beauty.[36] He had a love of the outdoors and a philanthropic bent, but he also had a shrewd business sense and a talent for making money—along with a strong penchant for getting his own way. He became a financial giant, amassing a fortune first in the financial world of Wall Street and eventually moving into railroads later in his career. While Henry had spent many boyhood hours in the forests of the Ramapo Mountains, his adult life became centered in New York City. It seems doubtful he expected to become the prominent citizen of the Hudson Valley that he eventually became. It very well may have been an accident that his life took this significant turn.

3. Edward H. Harriman and his wife, Mary, upon their return from Europe, 1909. Courtesy of the Orange County Historical Society Archives.

1. Portrait of George W. Perkins Sr. Courtesy of the Palisades Interstate Park Commission Archives.

5. William Addams Welch seated at his desk. Courtesy of the Palisades Interstate Park Commission Archives.

6. W. Averell Harriman delivering a $1 million check to George W. Perkins Sr. as part of the Harriman donation, 1910. Courtesy of the Palisades Interstate Park Commission Archives.

C. G. H. Jr. was born Mar. 6th 1901. The reason for my absence.

Mrs. John Holland.

Dear Madam: —

At the Annual Meeting of the League for the Preservation of the Palisades, held at the United Charities Building, Thursday, February twenty-first, you were unanimously elected a Director at Large of the League.

Respectfully yours,
Katharine J. Sauzade,
Cor. Sec.

Englewood,
March second,
1901.

7. Handwritten note from Katharine Sauzade to Cecilia G. Holland regarding her appointment to the League for the Preservation of the Palisades. Courtesy of the New Jersey State Federation of Women's Clubs Archives.

3
Wall Street and Railroads

It was a September day in 1886 when thirty-eight-year-old Edward Henry Harriman (Henry to his friends) accompanied his boyhood friend Edward Parrott to the Goshen Courthouse in Orange County, New York, for an auction to liquidate the Parrott family estate.[1] The auction marked a long run of bad financial luck for the Parrotts. Henry, on the other hand, was at a point in his life where he was quite comfortable financially. He was the owner of a successful investment firm on Wall Street and was just beginning to dabble in the railroad industry, an industry within which he was to build an empire that would make him a very wealthy man. That September day he had traveled from the east side of Manhattan, where he lived with his wife and children, to Goshen, in the Hudson Valley, with the simple objective of supporting his friend and observing the presumably humdrum legal proceedings. In short order, however, he found himself among a group of enthusiastic bidders. It had become apparent that these were timbermen bidding on the Parrott land to harvest its forests. Well acquainted with the Parrott estate from his youth, Henry was appalled at the thought of the most beautiful woodlands he had known being stripped of their most essential resources. He decided he had to do something.

Thus, he found himself raising his hand repeatedly, outbidding each timber king until he had acquired the entire parcel himself. When the auction was over, he was the successful bidder on nearly eight thousand acres of land in the Ramapo highlands, for the total sum of $52,500.[2] Sources conflict on whether this transaction was the source of a falling-out between Harriman and the Parrotts since his bid had come in below what the Parrotts had hoped to recoup for the property.

It is indisputable that, at least until that moment, the relationship between the Parrotts and the Harrimans was one of long standing. Henry's father, Orlando, had been a friend of Robert Parker Parrott (inventor of the Parrott rifle during the Civil War) and his brother Peter, who together managed several of the iron mines and furnaces in the hills of Orange County. The brothers subsequently purchased the mines and the surrounding acreage, amassing a substantial amount of wealth from the mining and manufacturing of iron ore and the production of the Parrott rifle. In 1877 Robert died, and Peter and his sons, Edward and Richard, took over the business. Within a few years, however, large deposits of iron ore were discovered in Pennsylvania—discoveries that had a significant impact on the marketplace and rendered the New York mines relatively unimportant. In the years that followed, the fortunes of the Parrott family steadily dwindled. By 1886 they were forced to sell most of the acreage they held, retaining only the family home and the land immediately surrounding it.[3]

Henry was successful at this point in his life, but he had not always enjoyed financial security. Although both of Henry's parents, Orlando and Cornelia Nielson Harriman, belonged to old and distinguished families, Orlando had chosen to enter the Episcopal ministry rather than pursue a commercial career.[4] This decision resulted in a constant financial struggle for the couple, which continued throughout their lifetimes. Nonetheless, they raised six children and managed to secure the essentials for the family. Henry, the fourth child of six, was born in 1848, in Hempstead, Long Island, where Orlando was serving as curate in a local parish. By the time Henry was old enough to attend school, the family was living in Jersey City, where he attended the local public school. As he grew, his parents began to recognize that he had a notable aptitude for learning and an easy understanding of complex ideas. They decided that he should be exposed to a higher-quality education. They enrolled him in Trinity School, across the river in New York City, confident that Trinity would provide him with greater opportunities for success in life.[5]

Although Henry was a good student and blessed with a keen wit and an impeccable memory, he was inclined to restlessness and was not always diligent. While there is little concrete information on Henry's early years, Harriman family biographer George Kennan wrote that school records indicate he was more inclined to fun and mischief than to serious study—ironically, however, he was consistently listed at the top of his class.[6]

Every morning Henry left his home in Jersey City and made his way to the ferry, crossing the river to Manhattan, and trudging another mile to the school, located on the Trinity Church property at the corner of Wall Street and Broadway. His route brought him into daily contact with boys from the street gangs of the west side of Manhattan. He was small for his age—even as an adult, he reached a height of only five feet, four inches—and thus would have made an attractive target for bullies, but he fared well. Potential antagonists did not count on his stubbornness or his ability to fight.[7]

At the age of fourteen, two years after entering Trinity, Henry declared to his father that he no longer needed to be in school; he thought it was time he made his way into the world. Orlando, of course, strongly disagreed, but Henry persisted, claiming that his early entry into the working world would benefit the entire family. After finally persuading his father, he quickly secured an entry-level position as an office boy and messenger in the established Wall Street firm of D. C. Hays and Co. Although he had family members in the lucrative mercantile and shipping industries, Henry was set on entering the fast-paced world of Wall Street. His diligence and enthusiasm paid off. It was not long before he rose from messenger to the ranks of "pad-shover"—the term for young men armed with pads containing current stock prices, which they provided to investors trading daily in the markets. In this preelectrical age, before ticker tapes, the pad-shovers provided timely information to investors who had to make on-the-spot decisions about buying and selling during the business day.[8]

As a pad-shover, Henry was on the trading floor daily, rubbing elbows with financial tycoons in action and being exposed to the inner workings of the financial world. The education provided did not go to waste. He watched and learned, building his skills slowly and deliberately, and by the age of twenty was manager of D. C. Hays and Co. Nevertheless, he was determined to open his own brokerage firm and create his place on Wall

Street. Within two years he secured a $3,000 loan from his uncle Oliver Harriman, a successful merchant, and used the money to purchase a seat on the New York Stock Exchange.[9]

At Hays and Co., Harriman had not only gotten a solid education on the ins and outs of the financial markets but also developed a broad network of acquaintances and patrons, many of whom were prominent speculators. Several of these men became his first direct clients.

Time passed and his business grew—as his commissions increased. While monitoring the markets for his clients, Henry also pursued occasional personal investments, consistently making money and never becoming overextended. With his reputation rising on Wall Street, he began to move into more exalted social circles, working with and befriending such financial icons of the era as Stuyvesant Fish, James B. Livingston, William Bayard Cutting, and the Van Buren family.

Charitable work—a social stepping-stone, then as now—was considered obligatory for someone in his position.[10] In 1876, along with his friend George C. Clark, another Wall Street financier, he became interested in the boys who were part of street gangs common at the time in New York City. He rented the basement of a local building, and, after recruiting several of his friends to help with the effort, he established the first meeting place of what was to become a vital community center for underserved boys in the city. Attendance was small to begin with but gradually increased, and by the early 1880s, the club was popular enough to necessitate expansion into the first floor of the building. The boys were given the space and equipment to play a variety of sports, but they also learned important life skills, ranging from how to fish to how to maintain personal hygiene. Harriman scheduled speakers to instruct the boys in various areas, and often the boys themselves would dictate the topics to be discussed. In 1887 the club was legally incorporated under the name of "The Boys Club of Tompkins Square" with control vested in a board of directors and an executive committee. In later years, the Boys Club began to offer summer camps in addition to its original activities. These camps were in the Palisades Interstate Park, in the hills of the Hudson Valley.[11]

Harriman continued to be closely affiliated with the club throughout his life, maintaining constant involvement with its concerns. In 1899, when the club was quite obviously outgrowing its quarters, he purchased several empty lots at Avenue A and Tenth Street. In 1901 he built a new complex for the boys at this location, spending close to $250,000 of his own money on the project.[12] He made an annual financial contribution to the club and was also known to cover financial shortfalls whenever the need arose. After his death, his family continued actively supporting the organization, with both of his sons serving on the board.

In 1877 Henry met Mary Averell, daughter of a prominent family from Ogdensburg, New York, who had interests in railroads and banking. She was in New York visiting relatives when she caught young Henry's eye, and they were introduced. The two were well suited to each other, and soon afterward they were engaged. On September 10, 1879, they were married in Ogdensburg at Mary's family home.[13]

Shortly after the couple married, Henry's new father-in-law, William Averell, appointed Henry to a seat on the board of his Ogdensburg and Lake Champlain Railroad Company and subsequently affiliated him with other Averell-owned railroads. Harriman was initially interested in railroads purely as investments, but he was soon captivated by the excitement of being directly involved in the dynamic, rapidly developing industry. In those days, railroads represented a new frontier, economically as well as technologically, and they were a primary focus for many on Wall Street. As usual, Henry was able not only to digest the financial technicalities of the industry but also to assimilate the subtle trends in operational strategies. He was intrigued by the idea of entering the operational world himself and had several ideas about how he might apply his financial and management acumen to increase efficiency and profitability. Intense and detail oriented, Henry saw railroads as a new world and one that presented him with unique opportunities to address a spectrum of new challenges. His instincts proved correct. Railroads would soon become the centerpiece of his career and the source of a vast fortune he would spend the rest of his life building.[14]

In 1881 Henry made his first independent venture into railroads. After forming a syndicate with his brother Willie, his friends Stuyvesant Fish and Sylvanus Macy, and Mary's brother William, he moved to acquire the Lake Ontario Southern Railroad, a small railroad in need of revamping.

He invested little capital. He simply changed the name to the Sodus Bay and Southern and marketed it to the Pennsylvania Railroad as well as the New York Central, two of the most prominent railroads in the Northeast. Its most valuable asset, an outlet on Lake Ontario, was attractive to both lines. Bidding ensued with both prospects, but the Pennsylvania Railroad emerged as the victor. Unsurprisingly, Henry made a handsome profit.[15]

By 1886 Henry had amassed enough money to buy the eight thousand acres of Parrott land. It would provide the foundation of his formidable estate. At the time of the auction, he and Mary had been married for seven years and had three children. The estate offered a vast and heavily forested retreat where the family could relax in the utmost privacy while luxuriating in nature. To Henry, his family was at the center of his world, and business trips were typically planned to include every member. The children traveled throughout the United States in the Harriman family's personal train coach and were exposed to international travel at an early age. The eldest child, Mary, born in 1881, was Henry's favorite. Her little brother Henry Neilson (nicknamed Harry) was born in 1883; he contracted diphtheria and died in 1888, before reaching his fifth birthday. Cornelia was born in 1884 and Carol in 1889, followed by W. Averell in 1891. The youngest, E. Roland Harriman, was born on Christmas Eve 1895.[16]

Immediately following the purchase of the Parrott land, the Harrimans began spending weekends in the large house on the property near Echo Lake. They decided to retain the name the Parrotts had given the estate, Arden, which had been Mrs. Parrott's family name.[17] Pleasingly, it was also the name of the forest in which Shakespeare's *As You Like It* takes place. The play is a pastoral comedy revolving around several main characters, one of whom, Orlando, was the namesake of Henry's father. In later life, the younger Harriman remarked, "Arden to me is the Arden of 'As You Like It' . . . , a retreat from the world of worries."[18]

In the spring of 1887, after a winter hiatus, the family resumed their weekend visits to Arden, traveling from their New York City home by railroad on a regularly scheduled train from the Erie Railroad station in Manhattan to the local rail station near Arden, and then by horse and buggy to the Echo Lake house in the valley. Eventually, as Henry's affiliations with several railroad lines developed, a private railcar for the family's use would be added to the regular run.[19] Arden seemed to have everything: woodlands, swamps, orchards, meadows, lakes, and streams, all teeming with wildlife. Less romantically, but more practically, it boasted a limestone quarry, several iron mines, numerous charcoal pits, two iron-making furnaces, sawmills, icehouses, access roads over the mountains, and numerous outbuildings and farmhouses. The environment gave Henry a respite from the intense and frenetic business world in which he operated daily. Not that he was bothered by such an atmosphere—on the contrary, his mind thrived on challenge and activity. Amid Arden's peacefulness, he threw himself into surveying the acreage, nurturing his forests, improving the roads, and developing the estate's drainage systems. Within a short time, he had hired a full staff of maintenance and construction personnel, and Arden became yet another project in his portfolio. He continued to amass additional parcels of land, and, in the two years between 1888 and 1900, he had accumulated about forty additional parcels that, when tabulated with the eight thousand acres originally purchased, amounted to nearly twenty thousand acres.[20]

As he became acquainted with the totality of his purchase, Henry soon realized that land and industrial sites were not the only things he'd purchased. Arden, it turned out, was a conglomeration of different elements in need of nurturing. It was home to a group of tenant farmers, five of whom specialized in dairy production. Dairy was the dominant offering, but other farms produced a selection of crops, including hay and corn, presumably used to feed the many animals needed to work the land.

Working closely with the tenant farmers, he moved to establish Arden Farms Dairy Company (which was a unification of the already existing dairy farms that had functioned as individual entities up to this point) and Arden Farms General Stores. He constructed a dairy barn outfitted with the most up-to-date equipment for milking and bottling. The barn boasted electric lighting fed by a power plant located on the premises that had been configured to service the barn and several other farm buildings, along with the Harriman family home on Echo Lake. By 1898 the dairy had been transformed into a unified business with additional milk contracts in place and new daily runs delivering dairy products to customers in nearby towns. Within a few years, the rapidly expanding routes served businesses and private customers throughout the county, including the towns of Central Valley, Arden, and Southfield, as well as the West Point Military Academy.[21]

Henry loved horses, and by 1888 he had added state-of-the-art stables, a stud farm, and a Standardbred (trotting horse) nursery to the estate. In a very short time, he emerged as one of the most prominent figures in the local breeding and racing community. He leased space at the Orange County Driving Park for his trotters, and the track he developed on the site became known throughout the state.[22]

Many of Henry's private accomplishments and attitudes involving his family and those individuals living on the estate paint an endearing portrait of the man, and, as far as I can determine, these stories are true. However, I must, of necessity, explain that much of the biographical information available on Edward Henry Harriman is provided through biographies written in the 1920s, a literary period characterized by flattering depictions of subjects. He was a controversial figure whose reputation varied widely according to individual perspective and, possibly, on whether a man found himself on the losing end of a business deal where Harriman was the winner, which often was the case.

Henry's actions in business suggest he was a firm believer that the "ends justified the means" and did not hesitate to employ methods others deemed questionable to accomplish his objectives. In the end, he was a powerful man, used to getting his way no matter what it took. As I dug for information on him, I was able to uncover several sources, from decidedly different perspectives. Some portrayed him as a ruthless businessman, but others who knew him in business recognized his intelligence, sharp memory, attention to detail, and ability to quickly analyze a situation. Whatever the observer's impressions, it seems he was typically not seen as endearing. Most characterizations describe him as brusque with a cold, impersonal manner. Many thought he lacked compassion. His "combative" nature and determination to get his way only reinforced this reputation. There were some financiers and railroad men of the time who felt he was mean-spirited and deficient in scruples and integrity.[23] One of his business partners, Otto Kahn, said of Henry:

> His was the genius of the conqueror, his dominion was based on rugged strength, iron will, irresistible determination, indomitable courage and, upon those qualities of character, which command men's trust and confidence, he was constitutionally unable either to cajole or disassemble. He was stiff-necked to a fault. It would have saved him much opposition, many enemies, many misunderstandings if he had possessed the gift of suavity. . . . I ventured to plead with him that the results he sought could just as surely be obtained by less combative, more gentle methods, while at the same time avoiding bad blood and ill feeling. Invariably, his answer was: "You may be right that these things could be so accomplished, but not by me. I can work only in my own way. I cannot make myself different nor act in a way foreign to me. They will have to take me as I am or drop me. This is not arrogance on my part. I simply cannot achieve anything if I try to compromise with my nature and follow the notions of others."[24]

Kahn was a member of the brokerage firm Kuhn, Loeb, and Company, and, along with Jacob Schiff, a partner in the same firm, he enjoyed a favorable relationship with Henry. Schiff (like Kahn, a member of New York City's German Jewish elite) had first met Henry in 1897 when they had competed to acquire the Union Pacific Railroad. He knew that Henry's Illinois Central was one of the best-run railroads in the country and the most profitable. Thus, when Schiff found himself in a bidding war with Henry for the Union Pacific, he was not averse to joining forces with him as a partner, even though Henry was viewed by some as a dangerous man with whom to do business. Schiff was a financier, not a "hands-on" railroader, so he gave Harriman free rein to manage the line and sat back, reveling in the resulting success.[25]

Henry's talent for alienating other businessmen gained him one of the most powerful financial tycoons of the time as an enemy—none other than J. P. Morgan.[26] Morgan and Henry had bumped heads several times in business dealings, and Henry often came out the victor. Morgan could not tolerate losing and was convinced his rival used underhanded tactics and hated him for it—especially when he won. In 1902, Morgan and his associate James J. Hill attempted to buy up the Chicago, Burlington, and Quincy Railroad to provide their Great Northern and Northern Pacific (NP) Railroads with a line to Chicago. At the same time, Harriman and Schiff also wanted to connect their Union Pacific Railroad with Chicago and decided to make a play for the CB&Q as well. The result was chaos.

Morgan had the advantage, and under other circumstances—that is, if Henry had not been involved—the opposing parties would have probably come to an agreement and negotiated a "community of interest" arrangement. It was a type of negotiation used at the time that usually enabled competitors to share a mutually beneficial interest in a venture. When Morgan realized his opponent was Henry, however, he flatly refused to budge. Henry, insulted by the reaction, was furious. He and Schiff went behind the backs of Morgan and Hill, buying up shares of Northern Pacific without their knowledge. Henry's attitude was that if Morgan and Hill blocked him from getting a piece of the CB&Q, he would go after its parent line, the Northern Pacific. After much back-and-forth, Henry and Schiff acquired most of the shares. The increased activity spurred the NP's stock prices to rise, which invited speculators to jump in and instigated a frenzy that overinflated the NP stock and the stock market in general. On November 8, 1901, Northern Pacific stock hit $1,000 and then tumbled into a free fall that took the market along with it. Some stocks dropped 50 percent or more that "Black Thursday," causing what was, at the time, the greatest financial panic in US history.[27]

In the end, no one won. Harriman and Schiff came out with more shares overall of NP stock, but Morgan and Hill held more shares of common stock.[28] The four men eventually realized, however, that a truce was in all of their best interests. Following a series of attempts between the parties to assuage the situation, an agreement was finalized. The news settled the market that Thursday; it was announced, and Northern Pacific closed that day at a reasonable level—up 165 points.

To provide a long-term solution to the battle for control of the CB&Q, Harriman, Schiff, Morgan, and Hill created the Northern Securities Holding Company as a vehicle for all the railroads under their collective control: the Great Northern and Northern Pacific as well as the Union

Pacific and the Chicago, Burlington, and Quincy. Henry was given a seat on the board and a vote in the company, a concession Morgan and Hill agreed to, albeit grudgingly.[29]

While the men may have avoided catastrophe for the entire country's financial market, their stunts fed the growing political sentiment that big business was out of control and restraints needed to be put in place to protect against such consolidated power.

With the assassination of President William McKinley and the subsequent ascension of a young politician named Theodore Roosevelt, the political atmosphere in Washington changed to one antagonistic to big business. Unlike his predecessor, Roosevelt was intent on enforcing the Sherman Anti-trust Act, which he argued was a tool to protect the common man from the greed and collusion of big business. As part of his program, he forced the Northern Securities Holding Company to eventually disband.

Over the next several years, Roosevelt engineered some similar confrontations between the government and big business. These altercations destroyed the relationship between Harriman and Roosevelt—who had been close friends up to this point—and in 1908 a permanent rift developed between the two. Their relationship could not withstand their political hostilities, and they never saw each other again.[30]

While E. H. Harriman was building his empire in railroads, George W. Perkins was making his name in life insurance, John D. Rockefeller had established the Rockefeller name in the oil industry, and J. P. Morgan was known for his steel mills and his railroads. All these captains of industry had established their place in history, and all had secured a fortune for themselves and their families. History tells us that when issues of public discord developed in their home region, they did not hesitate to contribute to the presumed solution. Donations of money, land, and often their participation were provided regularly.

At about the same time E. H. Harriman was purchasing that first parcel of land in 1886, at the Parrott auction, while he was building his railroad empire, many of his newly acquainted neighbors within the Lower Hudson Valley were joining together to organize their efforts protecting the valley from the rapidly encroaching quarries that threatened to destroy the century-old cliffs called the Palisades that bordered the Hudson River.

4
The Right Man for the Job

One such concerned new neighbor was prominent financier George W. Perkins Sr., who moved from Chicago to New York City in 1892. He and his family purchased a home (from the uncle of E. H. Harriman, no less) in Riverdale-on-Hudson, set high on a bluff on the east shore of the Hudson, directly across the river from the Palisades, and, while he was pleased with the location and the view, he was dismayed at the rock quarrying and the all-hours blasting taking place across the river. Before long he joined with his neighbors in the movement to stop the explosions and save the Palisades, which was fast gaining momentum.[1]

George Walbridge Perkins was a successful businessman and New York Life Insurance Company (NYL) vice president when he arrived in New York. Following several effective years in the Chicago marketplace, the company decided to send George to New York to develop the company's interests in that market. Not only did he succeed in establishing himself in the insurance market, but his prominence in Hudson Valley would grow far beyond that of a resident dissatisfied with quarry blasting. His destiny would include his appointment as the first president of the Palisades Interstate Park Commission (formed in 1900), along with the distinction of being a major philanthropist and community servant. His status within the New York banking community allowed him to accomplish great things for New York Life Insurance and the Palisades Interstate Park Commission. J. Pierpont Morgan would attempt to lure Perkins away from NYL to work for him in several capacities. His life would be a testament to hard work and community service, as he consistently placed benefits for the community above his self-interests.

Born in Chicago in 1862, the eldest of four children born to George W. Perkins and Sarah Louise Mills Perkins, "Georgie" (as he was called) had a simple childhood. His family considered him a warm and loving boy, but not one with above-average intelligence or strong ambition. He attended grade school but did not go beyond that because of his family's precarious finances and the fact that he disliked formal schooling.[2]

George's father held a position at a local insurance agency, which eventually merged with the New York Life Insurance Company, a development that provided many new opportunities for the elder Perkins (and later for his son) and at the same time a better standard of living for the family. Unfortunately, soon after George's father's advancement, his mother, Sarah, died giving birth to her fifth child, and her baby died along with her. Young George took his mother's death as a serious blow. She had been his champion and had recognized his strong points where his father had not. Willie, George's younger brother by two years, was always considered more likely to succeed owing to his dynamic personality, and, therefore, when the elder Perkins decided to bring the boys into the insurance business, Willie was the one recruited as his assistant, while George was placed on the general office staff as a clerk.

Letters written by George's father to him over the years indicate, in vivid detail, his father's intense focus on correcting George's supposed shortcomings. According to his father, George needed to fight against procrastination, develop diligence and attention to detail, organize his time better, and improve his behavior in myriad other ways.

In 1882, a fresh tragedy struck the family—Willie contracted typhoid fever and died. Four years later, in March 1886, the elder Perkins died of pneumonia, and twenty-four-year-old George was left as head of the family, responsible for the surviving family members. Without hesitation, George accepted his new role as guardian of the Perkins family legacy and set out to prove himself worthy of the role, both at home and at New York Life.

He was offered his father's position at New York Life temporarily until the company could decide how to proceed. However, when it came time to assign someone to the territory permanently, the position was given to a senior general agent, L. C. Vanuxem.[3] George was instead offered the state of Indiana, which, at the time, was a depressed area. While George tentatively agreed, unbeknownst to his company, he traveled to Wichita, Kansas, and subsequently to Denver, Colorado, both booming markets. He was soon writing hundreds of thousands of dollars' worth of policies that he forwarded directly to the New York office. By the end of the year, he had sold approximately $735,000 worth of insurance on his own, and through an association formed and managed by George, he brought in additional sales amounting to $2.4 million!

With this accomplishment in his pocket, George returned to the home office and negotiated for the Colorado, Utah, New Mexico, Idaho, and Wyoming territories. The company consented. However, rather than a commission-based contract that was standard for insurance agents, they offered him a salaried position in management. (The company was fearful that, based on his accomplishment, he would be due an exorbitant commission if he continued under his original contract.) The company told George that this arrangement would provide other incentives and rewards, including more opportunities for advancement.[4] At the young age of twenty-five, he had secured a position as manager, directly employed by the company. He was now a part of the corporate hierarchy and in control of operations in five states. His office would be in Denver, and in 1888 that is where he moved his family, which at this point included his stepmother, Emily; his sister Emily; and his brother Edward.

In 1884, while still in Cleveland, George was introduced to Evelina Ball and was immediately smitten. However, it took several years before the two decided to marry.

They were married on August 6, 1889, and left immediately on a combination honeymoon and business trip that had them crossing the western part of the country several times over; it was nearly a year before they finally settled down in Chicago. George's territory and responsibility for New York Life had grown significantly, and he now supervised more than 350 salesmen from his small office in Chicago's Rookery Building. His sales force was spread over several states, and he was constantly on the road visiting agents.

Amid several scandals that threatened New York Life in the 1890s, George assumed a leadership role within the company and managed the damaging publicity, ultimately preserving the company's reputation and, in some opinions, its very existence. Despite his successful efforts on behalf of the company, he was unable to save the sitting president's job, and ultimately William H. Beers, the president of New York Life Insurance Company, was forced to resign on February 8, 1892.[5] George's diligence in personally taking on the role of public relations for the company during this trying time elevated his status yet again to another level within New York Life.

Within days of the president's resignation, John A. McCall, a former Equitable employee, was named as replacement.[6] Many employees resisted the selection of McCall given that he came from one of New York Life's harshest competitors, but McCall announced that he would appoint George as vice president, essentially putting him in charge of all agency matters, and this decision resulted in bringing the trustees together to approve McCall's appointment. For George to have risen to vice president of a major insurance firm by the age of thirty was no small accomplishment. However, there were certain downsides. The new appointment dictated that he relocate his family from Chicago, a city that both he and Evelyn (as he called her) loved, to New York City, an unknown environment, and he was understandably apprehensive. But in the end, he embraced the adventure before him and set off with his young family in tow. In 1891, Evelyn had given birth to a daughter, Dorothy, whom George idolized. While George truly enjoyed his position with NYL, there was no doubt that throughout his life his family was his priority.

By 1892 George W. Perkins was a force to be reckoned with. George, Evelyn, and Dorothy arrived in New York and settled into their new home in Riverdale-on-Hudson, a quaint, picturesque town set high on a bluff overlooking the Hudson River. The property had a majestic view of the Palisades directly to the west, across the river. The family renamed it Glyndor, a combination of their three first names: George, Evelyn, and Dorothy.[7]

Not too long after the family moved into their new home, the company decided that George should travel to Europe to audit the New York Life holdings in those markets. Together with Evelyn and Dorothy, George set out to explore the mysteries of the Continent.

McCall had specifically requested that George review the organizational structure of the European offices. In typical fashion, he jumped at the challenge, and it wasn't long before George identified some obvious problems. He put a plan in place that promised to increase the revenue and efficiency of the branch offices. While overseeing the plan, George continued to travel back and forth across the Atlantic. With each trip, his understanding of the rules and regulations of each foreign territory deepened, and he developed sound relationships with valuable colleagues across the Continent. In 1895 the government insurance regulators in Switzerland, Austria, and Germany—all countries where New York Life was active—decided to reset the standards for foreign insurance companies doing business in their domain. Most American companies could not meet the new standards, making it impossible to continue doing business. The American companies (New York Life being one of them) were powerless to fight back. Their operations in these countries ceased. In 1897 George found himself back in Europe again. Owing to George's focused efforts on revising certain company processes, the Austrians reconsidered licensing for New York Life. In 1898, after meeting with contacts within the Swiss Bureau of Insurance and agreeing to adjust several of NYL's practices, he announced that the Swiss had also agreed to grant New York Life authorization to do business. Things in Germany, however, would not be so simple, but George was not ready to give up just yet. He devoted a significant amount of time to developing relationships in Germany and learning exactly what the objections were to American insurance companies. The adjustments were a bit more complicated and the Germans a bit more demanding, but again, because of his astute negotiation skills, he ultimately was successful in reestablishing approvals with the government of Germany, just as he had with Austria and Switzerland.[8] He then moved on to Russia, where he continued his streak of success. George Walbridge Perkins had single-handedly opened huge new markets for his esteemed company. Upon his arrival home, he was deservedly received as a hero.

In 1895 Evelyn gave birth to their second child, George Jr., and by the turn of the twentieth century, George had risen in acclaim to be recognized as the number-two man at New York Life. The firm had grown from the smallest of the three insurance goliaths to the largest, having reached a billion dollars in revenue, with much of the credit owed to George W. Perkins.

While George was distracted by the problems of New York Life in Europe, the issues regarding the Palisades in his backyard had only worsened. When he arrived in Riverdale-on-Hudson, the destruction of the Palisades cliffs across the river from his home was starting to be recognized as a serious threat to the valley. Within two years, the cause of Save the Palisades had grown to become a semiorganized effort supported by virtually all the residents. The movement received a large amount of publicity, with articles appearing in the newspapers with relative frequency. It had become a cause célèbre of the time, with prominent citizens and politicians enmeshed in attempts to find a solution. Several quarry companies had bought property along the cliffs to mine the existing basalt rock, and their business was booming. Basalt was a major component of the paved roads cropping up everywhere to accommodate the fashionable new mode of transportation quickly gaining popularity across the New York region and the country: the automobile.

George was disturbed by what he saw happening across the river, not because his young children, Dorothy and George, were both often awakened from their afternoon naps by the daily explosions, but because he realized the consequences of this irresponsible attack on their environment. If this flagrant disregard for the pristine cliffs continued, the geological beauty that took aeons to form could be wiped away in a matter of years.[9] It was with this fear in mind that George took up the gauntlet as cocrusader with his neighbors, and together they began to formulate a plan of what could be done to stop this monster in its tracks.

By the late 1890s, America had industrialized at a breathtaking pace. At the same time, an appreciation for the natural beauty of the environment was slowly taking hold, although it remained a low priority in the minds of many. New York City, the country's foremost financial and industrial center, was located a mere forty miles south of the Hudson Valley area, and with the quarries in place and the constant blasting of the Palisades, the entire valley was poised to be rapidly transformed—that is, unless an intervention could be organized.

Preserving the Palisades posed a unique set of challenges. Since the cliffs spanned two states, the solution must be approached with a strong spirit of cooperation. Those persons involved would be undertaking an unprecedented challenge.[10] Although New York and New Jersey were not typically adversarial and usually exhibited neighborly collaboration when faced with a bistate issue, they were nonetheless two separate governments with different priorities and political factions; to get full agreement on any plan would not be easy.

Late in 1895 an article appeared in the *New York Times* in which the perplexing problem of the Palisades was again discussed at length by residents of several northern counties in New Jersey: politicians and their constituents. The city of Paterson's Democratic leader, prosecutor William B. Gourley, was quoted as saying, "I am very glad the New York Times has started to agitate this matter." He continued to state that saving the Palisades was a matter of utmost importance—one that every thinking man in the state would agree on. Several of his associates were also quoted, expressing their opinions and concerns. Judge John Hopper concurred, "Preserve the Palisades by all means." Judge James Inglis Jr. remarked, "I favor immediate action by our representatives in Trenton." Mayor Brann (of Paterson) expressed his great pleasure in reading an earlier posted article on the issue. "It was a magnificent picture in prose, calculated to stir the blood of Jerseymen to action at once." New Jersey state senator Robert Williams offered, "Well, you may quote me as being heartily in favor of any movement that will prevent the further destruction of one of the grandest and most beautiful bits of natural scenery in the United States." Those individuals quoted, for the most part, agreed that the remedy for the problem was to be found in legislation. Senator Williams added that he favored New Jersey, in cooperation with New York, acquiring, by condemnation proceedings, at least the front and slope of the Palisades: "This land could readily become a magnificent boulevard along the river's edge."[11]

On January 8, 1895, after delivering his annual message to the legislature, Governor George T. Werts of New Jersey appointed three members to a Palisades study commission. Within a month, Governor Levi P. Morton of New York also appointed three members to work with their New Jersey counterparts. In short order, the two-state commission came up with an idea, recommending that the section of waterfront on the Palisades stretching from Fort Lee, New Jersey, to Piermont, New York, be given over to the US government to establish a military reservation. Before the end of the year, a report was offered documenting the findings and opinions of the commission and presenting a course of action: "This Palisades mountain, sixteen miles long and two miles wide, on which there are not one thousand inhabitants affords the only practicable site on the Atlantic seaboard for a hundred miles or more, for a practice and maneuvering ground for troops of all branches of the armed services from all of the adjacent states, of sufficient extent and of such varied topography as to be eminently fitted for the purpose."[12]

If this proposal was successful, the Palisades would be protected by the federal government, and the arrangement would relieve both New York and New Jersey from the responsibility of finding a solution. However, when the bills were submitted to the Committee on Military Affairs of the House of Representatives in the Fifty-Fourth Congress in 1895, it concluded that the Palisades had little military value, and, therefore, Congress rejected the idea.

The two states refused to accept defeat. Three years later, with only minor revisions, the same bills were presented to the Fifty-Fifth Congress, but again, they failed to pass.[13] The matter of the military park failed both times it was presented before Congress because, in the opinion of many members of Congress, the property did not lend itself as a military training camp. Nor did Congress acknowledge any responsibility for solving a problem they felt belonged squarely to the two states involved.[14] And so, the quarry blasting continued . . .

8. Ironworkers who built the Bear Mountain Bridge posing on opening day, 1924. Courtesy of Frank Goderre Archives.

9. Portrait of George W. Perkins Jr. Courtesy of the Palisades Interstate Park Commission Archives.

10. Portrait of Howard Carter Baird, designer of the Bear Mountain Bridge. Courtesy of the Century Association Archives.

11. The Tench family, Douglaston, Long Island, about 1915, at a special celebration honoring Tench's mother, Ellen Murray Tench, widow of William Eastwood Carruthers Tench, then deceased. Frederick Tench and his wife, Emma, are third and second from the left, respectively. Courtesy of the Tench Family.

12. Portrait of E. Roland Harriman (cover photo, June 10, 1944, edition of *Finance* magazine). Courtesy of the Orange County Historical Society Archives.

13. E. Roland Harriman and his mother, Mary Harriman, on opening day. Courtesy of the Orange County Historical Society Archives.

5
The Original Jersey Girls

Those individuals who were searching for an answer to the problem continued their quest to save the Palisades. However, the cause appeared to be all but lost until a new champion emerged in support of the Palisades conservation. It was 1896 when the US Congress was reviewing the legislation from New York and New Jersey about a military training camp, and at the same time a group of women was meeting at an Englewood residence to discuss the formation of the Englewood Women's Club. The club was to be a new chapter of the New Jersey State Federation of Women's Clubs, which itself had been organized only two short years earlier, in 1894. The leaders of this new Englewood chapter were Elizabeth Breeze Vermilye, who would become its first recording secretary; Adaline Wheelock Sterling, its first president; and Sarah Sophia Dana Loomis, one of its first vice presidents.[1] The women involved in these organizations were interested in intellectual and cultural pursuits and issues of public significance. At this time in history, women still did not have the right to vote and were cautioned to stay at home and tend to their homemaking chores. Nonetheless, these women took on several difficult issues impacting their communities. Most women of the time simply provided support for their husbands, but the members of the NJSFWC were of a different ilk. Their awareness of the world around them led them to seek out issues that needed attention and work to improve those conditions. Sarah Sophia Dana Loomis was one of those unique individuals. Born on June 23, 1851, in Danville, Caledonia County, Vermont, to parents who recognized the value of education, Sarah's apparent intellectual curiosity was identified early on. In a time when few girls were given a secondary education, Sarah's father supported her in her wish to attend Vassar College in Poughkeepsie, New York. After graduating in 1873, she founded a school in Syracuse, New York, where she taught for several years.

In Syracuse, she met and married Chester Hicks Loomis, a well-known American portrait painter and landscape artist, and together they traveled to Europe and lived for a time in France. Their first son, Charles Dana Loomis, was born abroad, and upon returning to the United States, their second son, John Putnam Loomis, arrived. Sarah and Chester settled their family in Englewood, New Jersey, and Sarah continued to educate people around her. She organized classes in her home for women interested in study and discussion. Her dedication to the group was obvious, and attendees developed bonds with her (and each other) that would last throughout their lifetimes. Years later, a member of the study group, Elizabeth Cutter Morrow, wrote a tribute to Sarah, whom she considered a friend and mentor. It was published in the 1923 edition of the Vassar alumnae yearbook, to commemorate the fiftieth reunion of Sarah's graduating class; it read: "We are a group of women from Englewood, New Jersey, who have known the class (of 1873) through Sarah Dana Loomis. Twenty of us have met at her house every week for fifteen years for study and discussion. There were no formal rules for the class, except that we always made our leader sit in the same chair in the middle, that the waves of discussion might roll over her equally from all directions ... to help us understand ourselves and our world better."[2]

The ladies of Englewood were well acquainted with the problem created by the quarries since the Palisades were located virtually in their backyards. Their concern for the environment here in Englewood and regionally was paramount in their minds. They were determined to find a solution to the problem and save the Palisades!

While most historians will credit the likes of George W. Perkins Sr., Theodore Roosevelt, Foster M. Voorhees, J. P. Morgan, and John D. Rockefeller with successfully rerouting history and saving the destruction of the majestic cliffs, and to be sure these men played a key part and should share in the credit, the truth is that without the women of the New Jersey State Federation of Women's Clubs, none of the story that follows would have occurred.

Many women were involved in the organization and the movement, but Elizabeth Breeze Vermilye, Adaline Wheelock Sterling, Sarah Sophia Dana Loomis, Cecilia Gaines Holland, and Katharine Jordan Sauzade played key roles.

It is fascinating to follow these women as they made their way through a world where women struggled to be recognized as persons of value separate and apart from their husbands, yet they not only made their way but also forged a way for others. They were independent, strong willed, and intelligent, and they often accomplished feats that their male counterparts could not. As for the issue of the Palisades, they indeed found a solution where others had failed, yet never truly received the appreciation they were due.

In May 1897, the newly formed Englewood chapter had the honor of hosting its parent organization at the Lyceum in Englewood on the occasion of their third annual convention. A member of both the NJS-FWC and the Englewood Chapter, Katharine Jordan Sauzade delivered an impassioned speech, "The Preservation of the Palisades." "We cannot escape the disgrace nor the just censure of the civilized world if we permit by further neglect the continued defacement of these grand cliffs," she declared.[3] A round of enthusiastic approval erupted from the audience.

Katharine was an ardent advocate for the Palisades and spoke frequently at local venues to stimulate interest and support from the public. There is little known about her, other than that she was passionate about improving the world around her, and her efforts testified to that passion. She was born Katharine Jordan in New York in 1832, the daughter of Thomas Demilt Jordan and Julia Ann Jordan, and attended New York City's public school system. In 1853, she married John Sidoine Sauzade, a novelist, railroad financier, and real-estate speculator who was credited with naming the town of Haworth in Bergen County, New Jersey, when a railroad he was affiliated with first established itself there in 1872. The name Haworth was a tribute to the hometown of the Brontë sisters, whose novels he admired. He and Katharine had two sons, Thomas and Robert, and in the early years, Katharine concentrated on being a wife and mother and raising her boys. When her husband, John, died in 1879 at the age of fifty-one, the boys were both grown. Katharine turned her energies toward her community, joining the NJSFWC. Accounts of her efforts with women's clubs and her support for saving the Palisades appeared in several newspapers at the time. In the publication the *Forestry Advocate*, she states that one of the practical duties of women was "to conserve the beauties of nature."

Her speech the night of the convention at the Lyceum rallied the membership of the federation behind the cause to save the Palisades—so much so it prompted them to unite and form a new subcommittee under the federation's Department of Forestry to study the issue. Elizabeth Vermilye was appointed as chairperson of this new committee and was joined by Cecilia Gaines, the sitting president of the federation at the time, along with Adaline W. Sterling, Katharine, Sarah Dana Loomis, and several others. Their enthusiasm for the cause permeated their new campaign, as they proceeded to educate the public on the need for conservation: they wrote letters, published news articles, made public appearances passionately outlining the need for action, and lobbied their state legislators to join them in the cause.[4]

Through the state chapter, the women enlisted the aid of Andrew H. Green, chairman of the American Scenic and Historic Preservation Society (ASHPS), who shared the belief that the quarries must be stopped.[5] Over the years, Green was very active within the city government of New York, having held the position of comptroller of the Central Park Commission through 1870. He was also comptroller of New York City until 1876 and was involved in the consolidation of the five boroughs into the City of Greater New York. These affiliations and others afforded him enormous influence and credibility. The political relationships that the women were able to stir up resulted in a meeting with New Jersey governor Foster M. Voorhees. Although the governor agreed with the women since he had championed the cause of the Palisades himself several years earlier, he did not think it would be possible to bring about any change in the current political environment. It was a futile fight, he lamented; after all, as property owners, the quarrymen could do as they pleased on their land. It was a fundamental American right.[6]

The women continued their campaign, and as they did, they followed the newspaper reports on the legislation still being reviewed by Congress to turn the Palisades into a military holding camp. The women were hopeful that it would pass. The announcement that the Fifty-Fifth Congress had rejected the proposal just as the Fifty-Fourth Congress had done brought with it the realization that there was no alternate plan ready to be implemented.

The NJSFWC was a fledgling group and had no experience with the fight before them. Over and over, their attempts were rejected, and they were understandably discouraged. Recounting the situation many years later, Cecilia Holland wrote that the feeling at the time was, "The men had tried and failed, so what chance have we!?" Nevertheless, the women marched on, armed with an innocence perhaps fostered by ignorance and a determined idealism, searching for ways to accomplish the seemingly impossible. According to Holland, for inspiration, the women looked to Harriet Beecher Stowe, who, not too many years earlier, had written a book that some said provoked the Civil War and subsequently played a key role in the abolition of slavery—another undertaking that many thought impossible.[7] If Stowe could exert such influence on matters of great societal importance, then they, too, would continue to persevere. They would not give up.

Four months after Katharine Sauzade's momentous speech at the Lyceum, Andrew H. Green and Joseph Lamb, both members of the ASHPS, were invited by Harrison B. Moore, the vice commodore of the Atlantic Yacht Club and a well-to-do member of New York society, to join him on his yacht, the *Marietta*, for a trip along the Hudson River to view firsthand the destruction being wrought by the quarrymen on the Palisades. The vice commodore also extended invitations to about forty members of the NJSFWC, including Katharine Sauzade, Elizabeth B. Vermilye, Cecilia Gaines, Sarah Dana Loomis, and Edith Gifford, wife of John Gifford, the state forester of New Jersey. Mrs. Gifford was an NJSFWC member and the new chair of the NJSFWC's Forestry Committee.[8] The yacht set out on September 22, 1897, from the Battery in Lower Manhattan at about eleven o'clock carrying the ladies, along with Green, Lamb, and George F. Kunze, all respected members of the ASHPS. Also on board were noted members of the US Army: Adjutant General Stryker of Trenton and Colonel Loomis L. Langdon, both honored veterans of the Civil

War. The journey was enjoyable, as the vessel made its way north along the Hudson River. Vice Commodore Moore ushered his guests into the cozy dining room and presented a sumptuous luncheon, but conversation on the Palisades dominated the event, resulting in a unanimous call for an immediate meeting on the subject as soon as the meal came to a close. Cecilia Gaines, president of the NJSFWC, spoke: "It seems to me when we Americans sell the very ramparts of our Hudson to pave the streets, we lose something incalculably dear. We can't do much in legislation, but we can rouse public opinion. Men may legislate but women can agitate."[9]

As the *Marietta* sailed into proximity of the Carpenter quarry, all on board were witnesses to the destruction in progress. No doubt the presence of the ladies, presumably unannounced, was not a welcome sight to the Carpenter Brothers quarrymen who had their own opinions on the acceptable place of women—namely, at home with their children. It was certainly not interfering in the important work of the quarry about which they knew nothing. The men and women aboard the *Marietta* looked on with sullen disapproval as the midday blasting continued.[10]

Other sectors of the populace also joined the crusade, and the newspapers were on fire with reports of supporters rallying together in various circumstances for action on behalf of the Palisades. One such matter involved another quarry, Brown & Fleming, holders of a riparian grant that provided rights to the entity for property they leased along the front of the Palisades, near the boundary line of New Jersey and New York. The New Jersey Legislature had passed a law in 1895 that granted the New Jersey Riparian Commission authority to revoke any lease on property where destruction of the continuity of the Palisades was occurring. It was determined that blasting done by Brown & Fleming constituted a violation of this law, and the commission revoked the lease. The firm sued to have the action overturned. A newspaper notice reported that the matter was scheduled to come before the Supreme Court of New Jersey under a writ issued by Justice Van Syckle and served on John Payne, secretary of the New Jersey Riparian Commission.[11]

Another example of the passion and determination for the Palisades was exhibited when a *Brooklyn Daily Eagle* reporter published an interview with Professor Charles E. West, a scientist and prominent educator. Dr. West was indignant when the reporter mentioned the destruction. He expressed an admiration of the cliffs that he had held for the past sixty years and scorned the contractors who were at that moment reducing one of the grandest scenic features in America to "a dead pile of stone for street pavements." He put out a rallying call to the city's many institutions and organizations, such as the Brooklyn Institute, Columbia College, University of the City of New York, and the Museum of Natural History of New York City, to name a few, to join him and other honored members of academia, in fighting back against this annihilation of a thing of beauty.[12]

The 1899 election of Theodore Roosevelt as governor of New York provided another positive development. In Roosevelt, the conservation movement now had a champion in a position of power. The governor had long been a proponent of conservation across the country; as an outdoorsman, he had personally enjoyed the natural beauty of the Palisades. Urged on by Andrew H. Green, Roosevelt joined with Governor Voorhees of New Jersey, and, using the previously passed legislation on the Palisades in both statehouses as a guide, they appointed a new commission to study the matter once again and develop new recommendations for preserving the land. Each state nominated five commissioners to participate; Elizabeth Vermilye and Cecilia Gaines of the Women's Federation were appointed to the commission on behalf of New Jersey.[13] After several months, the commission presented a plan to have the two states jointly buy the 737-acre parcel of land along the base of the cliffs at the low-tide water mark, which would deny the quarries access to their land.[14]

In March 1900, the findings of the commission were memorialized in two identical bills, each of which was presented to the respective New York and New Jersey state legislatures for a vote. The bill in New York passed easily; the one in New Jersey did not.

The bill submitted to the New Jersey Legislature met with much more intense opposition since most of the quarry owners were in New Jersey. In the face of this opposition, Vermilye and Holland intensified their efforts and enlisted the NJSFWC membership to ensure the bill's passage. They wrote letters to local newspapers, scheduled speaking engagements before groups of local citizens, lobbied legislators, and spread information and enthusiasm on the topic. The bill was presented a second time, and this time it passed, without any exclusions, and was signed by Governor Voorhees.[15]

Included, surprisingly, was the "appropriation," which was a stipulation that lawmakers would be empowered to condemn and acquire land. This condition had been perhaps the biggest sticking point with everyone who opposed the bill the first time around. Also included was the provision to allocate $5,000 as New Jersey's contribution to a capital account for expenses.[16] The bill established the Palisades Interstate Park Commission that would manage the parkland acquired in cooperation with the state of New York. The first cooperative venture between two states to govern one area was now a reality.

Years later, Cecilia Gaines Holland would reflect on these efforts in a November 1930 article in the *New Jersey Bulletin*, a publication sponsored by the NJSFWC. The article was titled "The Saving of the Palisades." Holland had been elected president of the federation in 1896 and served through 1898. Her article read in part:

> The part that the New Jersey State Federation of Women's Clubs had in saving the Palisades of the Hudson is an interesting and inspiring thing for us to realize today. It was an enterprise where devoted patriotism and the sublime audacity of inexperience sought the conquest of the actual by the ideal, and by determination and persistence aroused public opinion and awoke forces that brought about the protection of one of the finest scenic wonders of our land, and checked the greed and vandalism which would have destroyed vast natural resources.
>
> I do not claim that women saved the Palisades. That would have been impossible. We were then not even citizens: we had no vote and no power, either political or financial. What we did have was enthusiasm for the shores of the most picturesque river in the world, and that love of conservation which has been always a characteristic of women. When we sang: "I love thy rocks and drills, thy woods and templed hills," we meant it!![17] (a stanza from *My Country Tis of Thee*, a patriotic anthem, lyrics written by Rev. Samuel Francis Smith in 1831).

Holland wrote, "I do not claim that women saved the Palisades," but, in reality, they did. Of course, they had the help of many—both men and women—but it was *their* determination that turned the tide in 1900. When the legislation was blocked and it seemed all was lost, it was *their* effort that made the difference and changed the outcome—and not once, but twice. Again, in late 1900 another bill seeking additional funding and the extension of the commission's powers would be met with resistance in the New Jersey Legislature, and again the women would rise and change the outcome.

The quarrymen had been furious at the brazenness of these women who thought they could interfere in a man's world—a world of business, industry, and profit—more specifically, *their* world. The women threatened to disrupt how they made their fortunes. It's probably not surprising to learn that they told the women in no uncertain terms to go back to their sphere[18] and tend to the matters of their home: cooking, needlework, raising children, and so on. Even men not involved with the quarries looked at them with skepticism . . . Why, these women didn't even have the right to vote! What made them think they were qualified to insert themselves into this matter?

Maybe they couldn't vote, but they had the passion and the astuteness to recognize the importance of the matter and, more important, the courage and persistence to act on it.

Cecilia Gaines was born in Elizabeth, New Jersey, on January 12, 1864, to Henry Gaines and Jane A. Gaines. She grew up and was educated in the schools of Jersey City, and after graduation she took a "grand tour" of Europe with her mother during which she studied at the Sorbonne in Paris. She returned to Jersey City and immediately became involved in her community, gathering friends to form the Old Volumes Book Club. Her ability to organize people to accomplish a goal singled her out among her friends. Soon after establishing the Jersey City Women's Club,[19] she was recognized as a leader, and in 1894, she was at the forefront of organizing the New Jersey State Federation of Women's Clubs. She served on various committees within the latter, most notably the Forestry Committee. By 1897, when the NJSFWC had become deeply involved in saving the Palisades, Cecilia and Elizabeth Vermilye were at the forefront of the movement.

Fellow federation members remember Cecilia's good sense of humor and appreciation of appropriate behavior and manners (though without the "stuffiness" that often accompanied these traits), along with her determination and strong beliefs.[20] In 1899 she married Dr. John A. Holland and continued to weave her many interests into the tapestry of her life; now, she would add being a wife.[21]

With the passing of the New Jersey legislation the second time around, the governors of both states signed their respective bills and proceeded to set in motion the structure of the new commission. The joint agreement gave the Palisades Interstate Park Commission jurisdiction over lands extending beyond the park and authorized its involvement in future planning and improvement, development, maintenance, governance, and management of the park as well as other surrounding properties, including state parks and historic sites in the Palisades region of New York and New Jersey. The governors of the two states were tasked with appointing five commissioners each to the newly formed commission, for a total of ten. Each commissioner would serve without pay.[22] Governor Roosevelt decided that there should be one commissioner designated to act as a leader, perhaps to be considered the president of the commission.

6
A Perfect Candidate

On an afternoon in March 1900, the phone rang in George Walbridge Perkins's office. Theodore Roosevelt, governor of New York, was calling to discuss the newly formed Palisades Interstate Park Commission. Perkins, an executive with the New York Life Insurance Company, was very familiar with the new commission, having been involved in the campaign that led to its formation.

Roosevelt called that day to draft Perkins as the new commission's first president. Caught off guard and faced with an already unyielding schedule, Perkins's first reaction was to decline. But Roosevelt was not taking no for an answer. He insisted Perkins consider the opportunity, and he proceeded to promote the cause of conservation with a long and impassioned soliloquy. Roosevelt's keen ability to recognize expertise in others had led him to identify George W. Perkins as the only acceptable candidate for the position, the governor explained. The man who would lead this new commission needed strong business acumen. He would have to be able to get things done in the face of opposition. He would have to be a man of the utmost integrity. In the mind of Theodore Roosevelt, that man could only be George W. Perkins.

Perkins listened to the governor and eventually relented enough to say he would at least think about it. He pondered the invitation, discussing the opportunity with his wife, Evelyn, and sought the counsel of those whose opinions he valued. Among the people he consulted was his close friend and colleague at New York Life John A. McCall. He telegraphed McCall and asked his opinion as to whether he should accept. McCall cabled back: "Palisades are about the right size for a monument to your levelheaded and successful work. . . . [A]ccept, of course."[1] Perkins accepted.

The challenge for the new commissioners would be steep.

The New York appointees were George W. Perkins, Nathan Barrett, D. McNeeley Stauffer, Ralph Trautman, and J. DuPratt White. The New Jersey appointees were W. A. Linn, Abram S. Hewitt, Colonel Edwin A. Stevens, Franklin W. Hopkins, and Abram De Ronde. Technically, each state could have had a ten-member commission, but it was decided and agreed that the states would combine their candidates, and the membership of each state's commission would be identical.

All ten took up matters regarding New Jersey at a meeting, made decisions, and adjourned. The same ten appointees called to order a meeting and addressed the New York issues. The process continued in an orderly and cooperative fashion. Unfortunately, although the New Jersey State Federation of Women's Clubs was a key player in the long and difficult road leading to the legislative approvals, no female representative affiliated with the federation was chosen to serve on the commission. The explanation for this glaring omission was something along the lines that the appointment of a woman might make the men involved "uncomfortable" in handling commission matters.[2]

The first official headquarters of the PIPC was Perkins's business office, located at 346 Broadway in New York City.[3] The first order of business was easily put down on paper, but not so easily accomplished: stopping the quarries from further destroying the cliffs. But even before this task could be addressed, essential administrative issues had to be handled. At the time of its founding, the PIPC had no office, very little money, and practically no staff (and, among those persons involved, no staff member had any real experience managing parkland). The legislation that authorized the creation of the commission included a pledge of some minimal funding: New Jersey approved $5,000, and New York approved $10,000. The PIPC treasury, therefore, held the grand sum of $15,000 to cover its operating expenses.[4]

As they rolled up their sleeves and got to work, the commissioners found that there was no current survey of the parcels of land within the commission's jurisdiction. Before they could even create a plan to eventually acquire these properties, they would first have to know how many parcels existed, where the boundaries were located, and the names of the legal owners. The commissioners decided it would be expedient to have an in-house engineer oversee these tasks and prepare a current survey. Several candidates for the position were interviewed, and the commission selected C. C. Vermule, a local engineer who was well regarded. This first official staff member was hired at a salary of $3,000 annually, to be paid from the commission treasury.[5]

The new PIPC engineer immediately set about completing the necessary surveys and gathering relevant information, along the way developing a system to organize and track the data involved. It quickly became apparent that determining the ownership of these various properties would be no easy task. The New Jersey properties alone included 147 parcels held by 112 different owners. Many of the properties were family owned, and, often, there had been no official transfer of title in generations, resulting in confusion and unclear boundary lines.[6] While Vermule was sorting out this mess, Perkins was busy creating a plan of operation for the new entity.

The first task for the new PIPC was to purchase the Carpenter quarry.

Perkins identified the parcel of land held by the Carpenter Brothers quarry—perhaps the most egregious offender when it came to the demolition of the cliffs. With its daily blasting, the quarry had caused a tremendous amount of destruction, and the PIPC was determined to put an end to it. When approached, the Carpenter Brothers gave an asking price of $200,000, but it was rumored that there was room for negotiation; Perkins offered $100,000. The Carpenter Brothers countered with $200,000, plus the promise of a $25,000 donation to be made to the PIPC once the sale was finalized. The parties continued back-and-forth negotiations on pricing around plus or minus $150,000 until Perkins finally secured a firm option of $132,500. A deposit of $10,000 was drawn from the commission's dwindling treasury to seal the deal.[7]

Perkins knew that the only way to obtain the balance of the purchase price was through private donations, and he immediately set about to secure them. His recent appointment to the board of directors for the National City Bank of New York had afforded him valuable contacts among the city's financial titans, and through one of his fellow board members, he was able to arrange an introduction to J. P. Morgan, one of the most successful financiers and industrialists of the day.

Morgan had a reputation for being a difficult man to deal with; he was often described as single-minded and stubborn. However, at the same time, he was known for his philanthropy, and he seemed to be agreeable to donating to the PIPC. "How much do you need?" he asked. Perkins told him he needed to raise the balance on the Carpenter property—approximately $125,000. Morgan agreed to donate $25,000, but he added that he also had an alternate offer. Unbeknownst to Perkins, Morgan had been following his career. He knew of his business accomplishments and, specifically, his success in getting authorizations for the New York Life Insurance Company to do business in multiple countries throughout Europe. He was impressed with Perkins and regarded him as an asset—one that he wanted for his own company. To that end, he offered to donate the entire balance of $125,000 if Perkins would leave New York Life and come to work for him.[8] Morgan had holdings in some of the largest industries of the day, including railroads, steel companies, and shipping interests. His empire was substantial, and from Perkins's perspective, this opportunity was not one to be taken lightly. It would be an impressive move upward for the diligent Perkins. Nevertheless, he was happy with New York Life, and he was not sure if the atmosphere at J. P. Morgan's company would be to his liking. He asked Morgan for more time to think about it, promising to get back to him with an answer shortly. After Perkins discussed the situation with New York Life, they offered to increase his salary to more than double his current compensation. Perkins declined Morgan's offer and stayed with New York Life.

On July 26, 1900, soon after the formation of the PIPC, George W. Perkins Sr. wrote to Elizabeth Vermilye, in response to a letter he had received from her on that date in which she expressed unmistakable consternation and concern. It seems a colleague of Perkins had contacted Vermilye to ask for her assistance: Could she suggest names of candidates to be considered for an auxiliary committee the commission was contemplating creating? Vermilye took the request as an affront to the work of her own recently formed League for the Preservation of the Palisades, of which she had been elected president. The women had created the league as a platform where they could continue to support and collaborate with the new commission since they had been denied a seat at the main table. As she explained to Perkins:

> I am already in close correspondence with most of the societies likely to cooperate with this work and have a promise that they will push its activity in the Fall. The object of the League is to spread information; stimulate interest; enroll those interested in Palisades' protection in an organized effort; and raise money to cooperate with the legally authorized Commission in its work of buying and laying out an Interstate Park. . . . You will pardon me if I say that it seems to me if this effort of saving the Palisades is to succeed it must be done by all those having its success at heart, working together. . . . I wrote a letter to the Commission asking for a recognition of the League, but have had no reply—although I think, as the creation of the Commission is principally due to the persistent efforts of the women, I am entitled to one.[9]

The frustration and dismay communicated by her words were unmistakable. Vermilye had led the crusade that was key to getting the legislation establishing the PIPC passed in New Jersey. Once the bill passed and the commission was formed, however, her efforts and those of the other women involved were overlooked by both state governors when it came time to name the park commissioners—because "some male members might be less free in their deliberations if women also served."[10] The women, to be clear, were excluded. In response, they formed the League for the Preservation of the Palisades as a way to continue to play a role in supporting the cause of which they felt so strongly. And now, she explained to Perkins, it seemed that once again their voice was in danger of being ignored, all while their support was being welcomed—indeed solicited—to organize a new entity obviously in conflict with their own efforts.

Perkins knew he had to correct this misunderstanding—and quickly. To that end, his response opens with: "My dear Miss Vermilye: I hasten to acknowledge receipt of your letter . . . , and to assure you of my appreciation of its contents." He goes on to explain, quite delicately, that he did not mean to encroach on the work of the league, and, indeed, it was precisely to avoid doing so that he had considered forming an auxiliary commission. He envisioned representatives from all supportive organizations participating at some of the PIPC meetings, thus ensuring that, as Vermilye suggested, all supporters of the Palisades would work together and the PIPC would not infringe on or duplicate work already being completed by others. He also assured her that her previous letter, which had not yet been acknowledged by Perkins or his colleagues, had been received. And it was brought before the commissioners for review on the same day it arrived. He apologized that an official response had not yet been sent out, explaining that the PIPC staff were still setting up their offices. But, he insisted, a letter would be in the mail before the beginning of the coming week—making her the first person to receive a formal response from the new commission. His closing reveals his high regard for Vermilye and the efforts of the league:

> Speaking as President of the New York Commission to you as President of the League for the Preservation of the Palisades let me assure you that I have but one desire and that is that the New York Commission and your League may in all respects cooperate in such a manner as to avoid all possible friction and work in the most harmonious manner possible for the one goal for which we are both striving. I shall esteem it a very great favor if you will at any and all times feel at liberty to address me in a perfectly frank manner, either in criticism of or by way of suggestion for the work of the Commission from time to time.
> Very respectfully yours,
> G. W. Perkins.[11]

After this correspondence, Vermilye responded with a letter acknowledging and accepting his explanation and graciously alluding to a productive relationship in the future.[12]

To persons who knew her, Vermilye's assertive exchange with the commissioner would have come as no surprise. She had long been a dedicated crusader supporting numerous causes and was not easily intimidated. She was used to standing up for herself and the causes she championed. Elizabeth Breeze Vermilye was the second daughter of the Reverend Ashbel

Green Vermilye and Helen Lansing DeWitt Vermilye. She was born in Newburyport, Massachusetts, on March 15, 1858.[13] Rev. Ashbel Vermilye had graduated from the Rutgers Theological Seminary and was offered his first parish in Little Falls, New York. From that time forward, he and his family moved from place to place following the various ministry opportunities offered to him. In 1882, he retired from the ministry and took up permanent residence in Englewood, New Jersey, dedicating himself to literature and community service. Elizabeth had followed her father's path of service early in life, and as a young woman, she often found herself involved in community causes. She served on the board of the Mary White Home for the Aged in Tenafly and was often welcomed in many local churches and community groups to speak on religious topics; one of her most well-known lectures was "The Bible as Literature."[14] Her subsequent affiliation with the NJSFWC and the Englewood Women's Club as well as her leadership of the League for the Preservation of the Palisades all demonstrate her energies and ability and her dedication to the critical issues of her time. Although women of the time were not afforded the privileges enjoyed by their husbands, fathers, and brothers, many strong-minded, intelligent women worked diligently to right the wrongs they identified and to create a better world. Elizabeth was one of them.

Perkins and Vermilye continued working together via their two organizations and developed a cooperative relationship. A few months after Perkins declined Morgan's offer of employment, Morgan invited him to breakfast, during which he again raised the topic of Perkins's employment. This time they worked out an arrangement agreeable to both: Perkins would come to work for Morgan to assist in creating several new entities, one of them being the US Steel Company, but he would continue part-time with New York Life. Perkins did not want to cut all his ties with New York Life, but, at the same time, he was attracted to the challenging work Morgan outlined.[15] The PIPC received an anonymous donation of $125,000, and the purchase of the Carpenter property moved forward. On December 24, 1900, nine months almost to the day that the commission was founded, the blasting at the Carpenter quarry stopped.[16]

Late in December 1900, the newspapers announced the cessation of the blasting at the Carpenter Brothers quarry. The PIPC continued to map out their course of action to quell all the trespassers who were endangering the environment within the valley. Although they were successful in this initial undertaking, there was much still to be done, and the commission was working on a very lean budget. The states were far from generous in their appropriations, and the costs of acquiring further property were great. They would need to supplement the funds provided by the government with private donations.

As the PIPC developed into an intricate participant in the valley, however, it would come to benefit enormously from the exceptional fund-raising capabilities of Perkins and his associates who ably cultivated the enthusiastic support of some of the richest men in the country, many of whom made their homes in the valley. When Perkins secured the PIPC's first donation—the $125,000 from J. P. Morgan—part of the agreement held that the donation would remain anonymous since Morgan wanted it to be used as leverage to "nudge" the state governments into appropriating more money for the new commission. New York agreed, but only if New Jersey would issue their appropriation first.

7
A Partnership Is Formed

Late in 1900, after Perkins had written the letter to Elizabeth B. Vermilye that smoothed out their misunderstanding, a second opportunity presented itself for the women of the NJSFWC to come to the rescue once again. A bill requesting additional money be allocated for the PIPC was put to a vote in the New Jersey Legislature. Just as the original bill creating the commission was blocked, this appropriation bill was also met with resistance. By this time, Perkins and Vermilye were on their way to developing a reciprocal relationship, and when Perkins reached out to her and the network of women's clubs for help, Vermilye sprang into action. Still active in the NJSFWC, and now also president of the League for the Preservation of the Palisades, Elizabeth alerted the women around her that they would need to summon all the persuasive power they could muster.

The league's membership included such members of society as Mrs. Ernest Thompson-Seaton, whose husband was a notable writer and naturalist, Mrs. Ralph Trautman, and Mrs. Frederick Lamb, both of whose husbands were considered distinguished residents of the Palisades who had served on commissions affiliated with the cause.[1] Another prominent member was Mrs. Washington A. Roebling, whose husband had built the Brooklyn Bridge (and would ultimately come out of retirement to oversee the cabling for the Bear Mountain Bridge twenty years into the future). To be clear, Emily Warren Roebling was much more than just Washington's wife, and to list her here as merely that would be a grave disservice to her and to history. While the Brooklyn Bridge was still under construction, Washington A. Roebling, the chief engineer of the project, was taken seriously ill and was confined to his home, not able to travel back and forth to the Brooklyn job site. His wife, Emily, stepped in and functioned as his right hand, allowing him to maintain control of the project. From his window in Brooklyn Heights, he had a clear sight of the bridge, and he learned the specific details of the project through daily updates that Emily provided him as she traveled back and forth from Brooklyn Heights to the river. Emily took direction from her husband and delivered messages to the workmen at the bridge. There was some speculation that as this arrangement progressed, Emily became her husband's engineering equal. An article appeared in the *New York Times* in 1883 that presented just such a presumption. It was assumed that after but a short period of study, she had taken on the duties of chief engineer—a comment that was not true. While her knowledge increased, she never rose to the level of an engineer, nor did she ever make that claim herself. However, she proved to be an invaluable asset to her husband, and, given the typical roles women of the day were assigned, her accomplishment was nothing short of outstanding.[2]

In answer to Perkins's second call for assistance on passing legislation, the women began a massive campaign of letter writing, public speaking, and lobbying their contacts. Once again, they were able to gain results. In March 1901, a bill was passed in the New Jersey Legislature providing the PIPC with an additional sum of $50,000 and reinforcing the right to eminent domain, just in case the commission was faced with difficulty in acquiring land. With New Jersey's commitment in place, a bill was drafted and submitted in New York and subsequently passed as well, awarding the PIPC an allowance of $400,000. Although there was some bickering over the disparity of the amounts, New Jersey offered to include the riparian rights to the New Jersey properties as an offset. With the additional $450,000, the commission was able to proceed with the acquisition of key parcels of land within their jurisdiction.[3]

In 1901, as Elizabeth Vermilye and the League for the Preservation of the Palisades were busy defining the organization's structure, Cecilia Gaines Holland was unanimously elected director at large of the newly formed organization. A handwritten note from Katharine Sauzade, dated March 2, 1901, informed Holland of her appointment to the league since she had not been present at the meeting a few weeks earlier. A personal comment in Holland's handwriting is visible in the margin of Katherine's note: "C.G.H. Jr. was born March 6, 1901—*the reason for my absence*."[4]

Cecilia missed the meeting owing to the arrival of her daughter, born within days of the meeting. Over the ensuing years, she continued to be involved in both the league and the NJSFWC. She was active in many of the individual chapters within the federation, mentoring new members, championing new causes, as well as addressing challenges presented in her work with the league.

Perkins was immersed in his work of growing the PIPC while, at the same time, E. H. Harriman was buying up parcel after parcel of land in the valley, enlarging his estate to a size of twenty thousand acres. His aim was not only to create an idyllic getaway for himself and his family but also to preserve the land from the atrocities occurring in the area. To help steward the land, Harriman hired countless contractors to address the many maintenance issues involved. At one point, he sought the services offered by the Yale School of Forestry, established in 1900,[5] to ensure that requisite maintenance was performed throughout Arden's forests. North of the Harriman estate was the US Military Academy at West Point. By the early 1900s, it had also begun implementing newly devised conservation practices on its sixteen thousand forested acres.

In 1902, after the dissolution of the Northern Securities Holding Company following the fight to take over the CB&Q Railroad by Harriman, Schiff, Morgan, and Hill, E. H. Harriman sought to bury himself in a new project at Arden. Late that year he retained the prestigious New York City architectural firm of Carrère and Hastings (designers of the New York Public Library) to develop a plan for a grand house at Arden for himself and his family.[6]

By June 1904, crews began felling trees in anticipation of breaking ground for the mansion on Tower Hill, the spot within Arden selected for the house. At the same time, construction began on an incline railway running from the base of Tower Hill to the house site, some 1,310 feet above sea level. The railway transported men and materials up to the construction site during the project, and after the mansion's completion, it would provide access for the family and guests. Preparations continued throughout the following year, and in April 1905 ground was broken for the mansion itself.[7] Much of the stone used in the residence was quarried on Arden land, and skilled masons—immigrants from Hungary, Italy, Germany, and Czechoslovakia—were hired to do the installation.

By mid-August 1908, several bedrooms and bathrooms, as well as a small section of the servants' quarters, had been completed.[8]

The two-story main structure boasted several towers, an observatory, a ballroom, and a dining room with a beamed ceiling and wooden wainscoting. From the observatory, the view included not only the immediate grounds but, more distantly, fountains, valleys, streams, and three lakes. As the work continued over the following year, it became clearer just how much the design had integrated the house into the natural scenery. The result was breathtaking. In June 1909, a friend of the family told the *New York Tribune*, "The house is simply magnificent, and magnificent because it is so simple."[9]

After the Carpenter quarry was closed, other quarries realized that doing business within the jurisdiction of the PIPC was going to become increasingly difficult. Slowly, more and more quarries moved farther north along the river, outside of the PIPC's authority. The commission had no power outside of the area defined by law as their jurisdiction, and since many of the problems simply relocated to unrestricted areas outside of their control, they lobbied the state of New York to pass new laws extending their authority. In 1906, with the passage of Chapter 691 of the Laws of New York, the jurisdiction of the PIPC was extended to Stony Point and lands along the west bank of the Hudson River in Rockland County, including Hook Mountain.[10] By 1907 the PIPC had acquired 521.13 acres in total, having spent $475,534.80 to acquire them.[11] This land would be incorporated into the parklands governed by the PIPC. In 1910, legislation was approved repealing an act passed in 1909 and amending the Palisades Interstate Park Commission's original act of 1900. This new legislation expanded the PIPC's jurisdiction once more, this time to include Newburgh and the Ramapo Mountains.[12]

As their jurisdictional authority grew, the amount of work commensurate with it grew as well. The administration of these properties increased, as did the negotiations and navigation of the legal snafus involved in their acquisition.

Sometime during the period between 1905 and 1906, a request addressed to the commission was received at their office. The request was met with reluctance and ultimately declined. Mrs. E. B. Miles, a member of the NJSFWC, had written J. DuPratt White, asking that, considering the many years of support proffered by the federation, the PIPC join them in securing a site on the Palisades within the park to be used to honor and memorialize the women who had fought to save the cliffs. The PIPC at first responded with a nonanswer, saying that it "was not in a position to take official action in that direction."[13] If the commissioners assumed that these women would drop the subject, they were sorely mistaken; the record details a flurry of back-and-forth "discussion" on the topic. The federation was the same organization that had fought and won the fight that produced the very commission they were now being rejected by once again. They would not go away quietly. The women instead proceeded to raise money on their own. They ultimately collected $1,507.69 through the NJSFWC and $1,531.70 through the League for the Preservation of the Palisades, for a total sum of $3,039.39.[14] They intended to use the money to purchase a site of their choosing and donate the title to the commission if that's what it would take to create a memorial within the park.

One of the PIPC's original commissioners, Abram De Ronde, joined the women in their petition. He decided that the commission was not acting in good faith by making the women use their own money to purchase a location for their memorial. He approached Perkins, J. DuPratt White, and the other commissioners and suggested that the PIPC should not only support the idea but also provide the land, thus allowing the women to use the money they had collected solely for building a memorial as soon as possible. He pointed out that the actions the NJSFWC sought to memorialize were already five years old. The board ultimately agreed with him, and the women were notified of the decision. The federation turned over its collected funds to the PIPC, and the parties entered into an agreement, outlined in a letter dated March 4, 1908, written by White to Mrs. Henry H. Dawson, president of the NJSFWC. According to the agreement: "The Commission is willing to undertake to have conveyed . . . a strip of land formerly forming a part of the Greene estate, such plot to extend from the edge of the Cliffs westerly to the Boulevard, a distance of about 1200 feet and to be in width north and south 100 feet, thereby comprising approximately three acres. . . . It is the intention of the Commission to have this site properly designated by means of monuments or otherwise, and properly mapped."[15]

Despite this small forward movement, however, the construction of a memorial did not start immediately, and it would not move forward for many years.

For now, the women accepted the promise and patiently awaited the next step. Elizabeth B. Vermilye continued her work pioneering causes that she felt needed her support, serving as chairperson for the Committee on Forestry and for the Protection of the Palisades, formed in 1899, within the NJSFWC, and, of course, as president of the newly formed League for the Preservation of the Palisades. Additionally, she became involved in the formation of the Bergen County Historical Society, an organization whose mission was (and remains) to collect items pertinent to the natural history of and papers related to the civil, political, military, and general history of Bergen County and adjoining counties in New Jersey, including Rockland County in New York, just over the New York–New Jersey border. In April 1902, she was sworn in as one of seven vice presidents of that society. Elizabeth's fellow member of NJSFWC and the Englewood Women's Club Adaline Wheelock Sterling was also one of the society's founding members and served on the first executive committee.[16]

Adaline was born in Brooklyn, New York, in 1847, one of eight children. She lived her life in and around Brooklyn and Englewood, New Jersey. She was a unique personality of the times, a woman educated beyond grade school; she attended and graduated from the Brooklyn Heights Seminary and then went on to create and administer a preparatory school for young women located in Englewood, thereby ensuring education for other young women. In 1899, she was elected president of the Englewood Board of Education—another accomplishment rarely attained by women at the time. Active in the NJSFWC, Sterling was a driving force in the creation of the Englewood Women's Club, along with Vermilye, Loomis, and Sauzade, and she served as the first president of that organization. She worked tirelessly within the NJSFWC throughout her life while maintaining involvement in many other community causes, most prominent among them education and voting rights. She also helped to found the Englewood Hospital and in 1888 served as president of their board.

The outlandish idea of granting women citizenship and the right to vote was initially introduced by a group of women in upstate New York in 1848. There were four women, headed up by Elizabeth Cady Stanton, who met on a hot July afternoon to enjoy a cup of tea together and ended up passionately pouring out their discontent to each other regarding their lack of rights: they were not citizens; they could not vote, and therefore were prohibited from participating in their government and the formation of laws; they had no property rights, and only limited rights for education and employment; and a myriad of other injustices. This tea party escalated into the first Women's Rights Convention and ultimately into a movement that would sweep the country. After seventy-two years and the involvement of thousands of politicians, organizers, administrators, and activists, women were granted the right to vote, an outcome that women agreed was not an end, but rather a beginning. This accomplishment would allow them to achieve other much-needed reforms.

Adaline Wheelock Sterling was a crusader. Her portfolio of accomplishments was impressive. Besides her achievements in education and her activism within the NJSFWC, she became involved in women's suffrage rights circa 1910 and later was installed as a leader in the Woman's Suffrage Party (Nineteenth Assembly District) of New York City. She served as corresponding secretary of the party as well as a member of the Women's Suffrage Study Club and the Women's Political Unions of New York City. She actively participated in the New York campaign of 1917 that won the women of New York the right to vote. She was a dedicated member of the Daughters of the Revolution (DOR, not to be confused with the Daughters of the American Revolution, or DAR), from 1894 until her death; and she was elected president-general of the General Society of the DOR from 1900 to 1904.[17]

Besides her activism, or perhaps because of it, she was also an accomplished writer. She penned many book reviews for *Harper's Bazaar* in the late 1890s, and by 1914 she was listed as editor in chief of the *Patriot*, a quarterly magazine put out by the National Society of the Daughters of the Revolution. She was the assembly district editor of the *Woman Voter* (a publication put out by the Women's Suffrage Party) from 1913 to 1916; from 1916 to 1917, she served as associate editor, during which time she wrote several updates on the activities of the Women's Suffrage Party of New York City. In 1922, she wrote *The Book of Englewood*, a history of that town, which is still in print today.[18]

8
Progress and Development

As part of the original legislation approved in March 1900 that created the Palisades Interstate Park Commission, the states of New York and New Jersey authorized the PIPC "to provide for the selection, location, appropriation, and management of certain lands along the Palisades of the Hudson River for an interstate park and thereby to preserve the scenery of the Palisades."[1] Originally, this mission statement included acquiring lands along the Palisades within the original jurisdictional boundaries. Between 1900 and 1909, the commissioners devoted their efforts to developing the cliffs and all the shorefronts along the Palisades. By 1909 this task was accomplished, and the result became known as the Palisades Interstate Park. As the PIPC evolved, its mission also advanced, and as it acquired large parcels of land throughout the valley, a system of parklands was developed to support and extend the original plan.

The commission soon gained the respect of the citizens in and around the Hudson Valley owing to the professionalism exhibited. Not only were they successful in acquiring the originally targeted properties, but as their reputation grew, donations of both land and money also increased. The list of private donors included J. P. Morgan and John D. Rockefeller, and later added William K. Vanderbilt, James Stillman, William Rockefeller, and V. Everitt Macy, among others.[2] Of course, from the beginning the PIPC's biggest patron was Perkins himself, who routinely donated funds and land, in addition to his time, expertise, and passion, all of which he provided without compensation.

On September 27, 1909, the Hudson-Fulton Celebration was convened. The PIPC had planned for more than four years to make this an event of monumental importance for the valley. The intent was to celebrate the history and the scenery of New York and at the same time formally dedicate the Palisades Interstate Park. Mr. and Mrs. Hamilton Twombly had donated sixty acres of land just a week before the event, and their donation completed the Palisades waterfront in New York and New Jersey as part of the park. The celebration did indeed include the dedication of Palisades Interstate Park, along with Bear Mountain Park, Inwood Hill in Manhattan, Verplank's Point near Peekskill, and dozens of smaller municipal parks throughout the region.[3]

The festivities kicked off with an impressive naval parade led by replica ships of Henry Hudson's *Half Moon* and Robert Fulton's *Clermont* sailing through New York Harbor, up the Hudson to Albany. The entourage included five naval squadrons totaling eight hundred vessels: yachts, steamers, tugs, tenders, gunships, submarines, and others representing US ships as well as foreign fleets.[4] The enthusiastic spectators numbered close to one million, and they witnessed not only this event and others on water but many events on land as well, all commemorating historical achievements in technology and engineering in New York State. There was even a demonstration of airplane flights by Wilbur Wright that amazed and entertained all who were present. It was a time of aspiration and hope; the public mind-set was changing concerning many topics. Those individuals who saw the utilitarian value of natural resources often clashed with those persons who felt these resources should be left in their unspoiled, purely aesthetic state.

Between 1905 and 1924, three very significant engineering achievements that would test the geological resistance of the locale were announced. All three were ultimately successful, but each in their turn faced resistance and concern from the people, who understandably feared the impact on the scenery and their daily life.

The first was a humanitarian requirement: with the fast-growing population of New York (at the rate of one million per decade) came the need to increase the water supply to New York City. The tunnel of the Catskill Aqueduct was the most difficult project of the three, but at the same time, it is the project that utilized the most up-to-date advances in engineering of that time. The plan was to flood twenty-four square miles of the Catskill Mountain watershed, encompassing two Catskill Mountain streams, the

Schoharie and the Esopus. The plan called for two reservoirs feeding into a central aqueduct system, which was designed based on the principle of gravity. The components were engineered so that the grade level of each would cause the water to flow downhill on its journey to New York City. There would be no need for pumps in this system—a system that would be part of the longest and most expensive aqueduct in the world. The capacity would exceed five hundred million gallons per day. The plan was approved, and by 1905 the project was under way. An accurate analysis of the geological terrain of the Catskills and the Hudson Valley was necessary to map out the most efficient and least costly route for the tunnel, beyond a doubt, the most challenging part of the project. The feat took twenty years to accomplish, but the expertise, ingenuity, and perseverance of the engineers prevailed. By 1917 the first phase of the Catskill Aqueduct system was completed.[5] The project was an accomplishment of monumental success. The problems encountered while crossing the Highlands gorge as well as the advancement of the processes utilized provided a wealth of engineering knowledge and benefited projects that followed for years to come.

While the construction of the Storm King–Breakneck tunnel within the aqueduct system left almost no visible evidence on the landscape, the other two engineering projects—the Storm King Highway and the Bear Mountain Bridge—both left unmistakable marks on the valley. Each was highly visible, and each had an impact on the surroundings in its own way.

Storm King Highway had been talked about since 1903. The only route from Cornwall to West Point and Highland Falls before the highway was established was up and over the mountain via a long and twisting, circuitous roadway. The new road would cut the trip in half and would link to the expanding highway system running between New York City and Albany. The project met with many obstacles, resulting in years of delays, but once the Palisades Interstate Park Commission became involved, the project gained support. Approvals were granted in 1914, but the inclusion of the United States in World War I in 1917 delayed the start of work until 1919. From the start of construction to its completion, the work spanned three years. The completed roadway was met with enthusiastic approval when it finally opened for public use in 1922. Not only did it cut travel time, but since the traveler could enjoy views of the Highlands from an elevation of four hundred feet, it soon became known as one of the most scenic experiences in the nation.[6]

By the time Storm King Highway opened, the campaign to build the Bear Mountain Bridge was already under way. Almost immediately following the highway opening, ground was broken, and bridge work began. Its success personified the community spirit present in the valley. Prominent in that campaign was the Harriman family, specifically Roland Harriman, who worked with Frederick Tench, the contractor who brought the project to them. Roland devised a financial program that raised the necessary funding. Owing to their economic power, a strong sense of philanthropy, and civic responsibility, along with shrewd instincts, the family has maintained a continuous role in the story of the Hudson Valley, the Bear Mountain Bridge, and the surrounding state parks. The western approach of the bridge sits on a Harriman parcel that was given to the state by the family long before the bridge was even thought of with the stipulation that it be for park use only.[7] Bear Mountain State Park is made up, in large part, of land donated over the years by prominent families of the region, including the Harriman family. Harriman State Park, Sterling Forest Park, and several surrounding villages also occupy land that was originally part of the Harriman family country estate.[8]

Besides the Harrimans, there were many other supportive families, among them the Rockefellers, the Perkinses, and the Vanderbilts. Together, the efforts of these benefactors brought thoughtful progress to the environment.

9
The Devil's Horse Race

Under the management of the PIPC, the Palisades Interstate Park became a popular recreational destination. Camping was very popular, and the number of people visiting the Palisades to indulge in this pastime increased as time went on. While the park's popularity was a positive development, it brought with it a new set of concerns, as the PIPC recognized the need to establish a system of regulations that would guarantee orderly and safe enjoyment by the public. J. DuPratt White, the commission's director of field operations, took on this challenge, creating basic regulations to quell the rising complaints and problems, the number of which seemed to increase in direct proportion to the increase in park patrons. As a measure of control thereby allowing the management to monitor how many campers were present on park property at any given time, White began requiring campers to apply for permits, though they were provided free of charge. He also came up with a set of standardized recommendations regarding the use of camping equipment in response to the constant inquiries that poured in from inexperienced campers. He introduced guidelines to regulate fire safety, accident-response and first-aid procedures, sanitation, and bringing food into the park. He established rules to address trespassing on private property adjacent to the parklands and prevent unruly patrons from misbehaving and disturbing other patrons.[1] While White handled the day-to-day issues involved in running the park and dealing with the public, Perkins continued to make land acquisitions and fund-raise.

Besides these operational challenges, White also had to contend with several administrative problems. For instance, the Township of Alpine persisted in presenting the PIPC with tax bills, even though no services were provided by the municipality. Neighborly discussion ensued in response but yielded no solution. In the end, the PIPC refused to pay, and the matter was dropped.[2]

Sometime in July 1907, J. DuPratt White wrote a memo to the other commissioners listing several pressing matters that needed attention. His list suggested a unique idea—one that had not yet been introduced anywhere in the country: he proposed a roadway be built to connect city to city, with stretches of land between them designed as "parklike" spaces, giving the traveler a serene, tranquil passage experience. This concept would become the seedling consideration for the Palisades Interstate Parkway that would provide a scenic ride through the Palisades, although it would not take form as an actual project until William A. Welch, who served as chief engineer and general manager of the commission from 1914 to 1940, launched the idea several decades later. In the interim, several "parkways" were developed throughout the country, such as the George Washington Parkway (planned in the 1920s, with the first section built in 1932); the Baltimore-Washington Parkway in Washington, DC (built during the 1940s); and the Blue Ridge Parkway (developed in the 1930s as part of FDR's New Deal) from Virginia through North Carolina. The Palisades Interstate Parkway would develop over several decades; construction would be delayed because of World War II. Over the years that followed, the PIPC faced local political resistance as they continued to fight for the approvals. It would be constructed in phases, not reaching full completion until 1958, when the last phase was finally opened to the public.[3]

Early in its infancy, the PIPC had to focus on organizing and structuring the agency and, of course, fund-raising to purchase land they wanted to preserve. As the agency grew and land parcels were acquired, the still fledgling commission had to get down to the business of developing these parcels as recreational areas for the public at large while maintaining the natural state of the lands. The commission was confident it could create the nature reserve the agency (and its donors) had dreamed of, but with the recent influx of assets came pressure from those individuals who expected things to be accomplished and changes to materialize—and quickly! Unsurprisingly, the logistics of making things happen were often complicated. Many ideas were introduced. Conditions varied, and methods to accomplish each could often inadvertently impact other projects.

And, of course, laws and regulations had to be considered. Given all of these factors, the commission decided it would need a designated person to execute the specific goals and aspirations of the agency—someone who could envision exactly what was needed and then make it a reality; however, that person must also be educated in the technical aspects of the job. Balancing all the demands and goals (some of which were competing) ensured that it would not be an easy undertaking. Perkins dreamed of preserving the natural beauty of the land, but at the same time opening it up to city residents and teaching them to love and protect it as well. The candidate he was searching for would have to share that dream.

Charles Wellford Leavitt was a successful engineer who maintained a thriving practice in New York City, handling projects in landscape design, civil engineering, and architecture. Sought after for many high-profile projects of the day, the firm was successful in landing contracts to design the Italian-style gardens at the Charles Schwab estate in Loretto, Pennsylvania; the J. A. Haskell estate in Red Bank, New Jersey; the gardens at the three-thousand-acre estate of James Buchanan Duke in Somerville, New Jersey; and many other prominent commissions. Leavitt had consulted with the Rockefellers regarding the gardens and approach roads for Kykuit, John D. Rockefeller's forty-room estate in Mount Pleasant, New York, but ultimately did not win the contract. The firm's catalog of famous projects also included several prominent racetracks, including Saratoga Race Track and Belmont Park, both in New York. Leavitt also ventured into municipal planning and worked on many federal and municipal parks. This work eventually brought him into contact with PIPC president George W. Perkins Sr.[4]

Leavitt would have seemed like an obvious nominee for Perkins's search. He and his firm had been hired by the commission for several projects, and his performance had seemed satisfactory. The PIPC held him in high esteem. However, around this time there seemed to be some controversy among a few of the New Jersey commissioners regarding Leavitt's performance on the Englewood approach road, a project that he was working on in conjunction with Alfred Nobel (an engineer famed as the inventor of dynamite, as well as for the establishment of the Nobel Prize). Because of certain problems, the commissioners wanted Leavitt removed from direct supervision of the road, although he would continue handling general engineering work for the PIPC. Therefore, there was some hesitancy in selecting him for the position Perkins had in mind.[5]

As Perkins continued his search for his ideal candidate, Hudson Valley was abuzz with the progress being made as the new Palisades Interstate Park Commission acquired parcels of land. Residents of the valley rallied in support of the work being done to remove those entities that might ruin the natural landscape. Well-to-do residents donated money and land, and others contributed their time and ideas to keep the movement alive.

In November 1907, Dr. Edward Lasell Partridge, a respected physician in New York City and a well-known resident of the northern slope of Storm King Mountain at Cornwall-on-Hudson, published an article in the *Outlook*, titled "A National Park on the Hudson." His concern for the preservation of the environment in and around the Hudson Valley was recognized throughout the region. His opinions were held in high esteem, and he was respected for his philanthropy. The article outlined his proposal for a national park to be created on a certain sixty-five-square-mile section of the Highlands.[6] This area had long been identified as the most scenic in the valley. Painters attempted to memorialize it, while sailors recognized its unique challenges. Within that stretch, the Hudson River passes through a narrow gorge and meanders between mountains carved out by glaciers centuries earlier. There the river gains speed, surging past West Point and onward beyond Bear Mountain. The early Dutch sailors called it the "Devil's Horse Race" because of the convergence of the tides, the currents, and the winds in such a way that a significant churning of the water resulted and made it extremely difficult to navigate. The surrounding land was owned by E. H. Harriman, also a passionate advocate for preserving the environment. Partridge and Harriman were acquaintances but did not often cross paths. However, sources note that the article was read by both Mr. and Mrs. Harriman and may have contributed to the unprecedented move by the Harriman family that would occur within the next two years.[7]

In early 1909, Charles E. Howard, superintendent of New York State Prisons, decided that Bear Mountain would be an ideal spot for a new prison stockade and planned to use his prisoners as laborers to build it.

When the state of New York announced its intention to carry out this plan and to do it close to the Harriman property line, E. H. Harriman reportedly went into a rage. The plan mentioned that two thousand inmates would be put to work quarrying stone from the mountain for road construction and harvesting lumber to build the prison structures.[8] Harriman, of course, did not relish the idea of putting a prison so close to his estate, but even more upsetting was the fact that a state agency was planning to defile the very land that he and his neighbors were trying to preserve—it was downright insulting. His thoughts raced. What solution could he devise to stop this plan? He decided that if he were to offer the state a gift—a sizable parcel of his land (a piece contiguous with the proposed prison site)—and stipulate that the land could be used only to create a park, it may serve to block this current proposition. Of course, he would suggest that the prison project be relocated elsewhere, preferably on the other side of the Hudson River.[9] Harriman immediately reached out to New York governor Charles Evans Hughes to discuss his plan. The governor seemed to be very interested in Harriman's proposal to donate ten thousand acres of land and provide a monetary gift of $1 million to the state of New York for the development of a state park, but he was unable to meet with Harriman before June 1, 1909, which was the day Harriman was to leave for Europe to seek treatment for his health.[10] Unconfirmed reports circulated that he was suffering from digestive issues, possibly cancer. Henry and Mary sailed for Europe on June 1 after notifying the governor and suggesting they meet when Henry returned. But by the time Harriman returned to New York, in late August, his condition had worsened. By this time, however, the Harriman mansion was almost complete—enough to allow Henry and Mary to move into their palatial new home.[11] The Harrimans settled in, and Henry continued to conduct his business, although he now did it from Arden, which prompted the newspapers to question the exact status of the railroad baron's health. The matter of the donation was set aside as a priority; Henry's wife, Mary, was aware of the correspondence between him and the governor, and she discussed the intended donation with her husband.[12]

As Henry moved through the myriad pressing matters sent over by his office, he sought to tie up a "loose end" he felt to be of importance. The tenuous relationship between Henry and J. P. Morgan had been adversarial throughout their lifetimes, from both parties. However, it appeared to be a matter that Henry wished to address when putting his affairs in order. Shortly before his death, Harriman asked Morgan to pay him a visit at Arden to clear up some business matters. The invitation must have had an urgent tone because Morgan responded without hesitation, taking a special train up to Arden without delay. The two men sat together for some time on the sunny veranda, and after resolving whatever business they had, they spoke about the grievances they each had harbored. Although we cannot be sure what specifics they spoke of, Maury Klein's biography of Harriman tells us that there were indications that both were able to let go of the history that pitted them against each other for so many years. The result was that each acknowledged understanding and good wishes for the other, and they shook hands as Morgan took his leave.[13]

Henry's enjoyment of the house was short-lived. He died there on September 9, 1909, one day shy of his and Mary's thirtieth wedding anniversary.[14] All construction work at the estate was shut down for a period of mourning, as were all locally owned Harriman businesses. A private funeral service for family and friends was arranged and included many New York City dignitaries as well as local villagers. The little church at Arden parish overflowed with crowds of employees, townspeople, and visitors from everywhere, who stood on a slope overlooking the burial plot at St. John's Cemetery. At exactly 3:30 p.m., every train and every workman on a Harriman line, including the Erie, halted for a full minute of silence in Henry's honor. Trains stopped, telegraphs were silenced, laborers dropped their tools, and office staff bowed their heads.[15]

Within the year following Henry's passing, Mary met with Governor Hughes, and, with the assistance of George W. Perkins, the president of the PIPC, they reached an agreement. Mary Harriman would turn over ten thousand acres of land along with a donation of $1 million, according to her late husband's wishes; however, she had several additional stipulations of her own. She required that the PIPC secure additional donations from other sources amounting to at least $1.5 million, bringing the total working capital to $2.5 million—a sum that the state of New York would then be required to match. This arrangement would provide a total of $5 million (roughly equivalent to almost $160 million in current-day dollars) to be used for additional land acquisition, necessary park development, or both. The meeting minutes of the PIPC's regular monthly meeting held on December 16, 1909, recorded these stipulations. Besides these important details, another major development impacting the commission's work was included: Perkins was nearing the completion of his negotiations with Albany to extend the PIPC's jurisdiction to include West Point.[16]

Once again, Perkins tapped into his fund-raising skills to collect the required $1.5 million in additional donations. Almost immediately, private donations began to roll into the PIPC. Both John D. Rockefeller and J. Pierpont Morgan pledged $500,000 each. The balance was the sum of generous donations from several valley residents, including Margaret Slocum Sage, probably one of the most influential women in history, who was a noted philanthropist and activist supporting conservation, women's suffrage, and women's education. She was the widow of Russell Sage, who had built a fortune in banking and railroads. After her husband's death, Margaret formed the Russell Sage Foundation with the $63 million estate he left her and began to distribute the money to charity. Her foundation was a prototype for many modern American charitable foundations. Two of her conservation-related projects were the creation of a seventy-thousand-acre bird sanctuary in Louisiana and the purchase of Constitution Island, a Revolutionary War historic preservation site located in the Hudson Highlands, which she gifted to the federal government.[17]

As the PIPC approached the $1.5 million goal, a follow-up meeting was scheduled for December 23. The minutes of that meeting show a detailed outline of the conditions of the Harriman donation, along with confirmation that the state of New Jersey would also contribute an amount deemed to be their fair share (subsequently determined to be $500,000). The obligation to relocate the prison site was reaffirmed, along with a decree that the PIPC's jurisdiction would be extended once more: northward along the west coast of the Hudson River to Newburgh and westward to include the Ramapo Mountains. This extension would provide the PIPC with the same powers it was granted in the original authorization of jurisdiction, which included the right to condemn land for roadway and park purposes. The following day, Perkins and White traveled to Albany to meet with Governor Hughes.[18] On January 6, 1910, Governor Hughes submitted the proposal to the New York Legislature for approval. At the same time, the New Jersey Legislature approved the state's share of $500,000.

The Harriman gift stimulated additional donations as well as state appropriations from both New York and New Jersey. In 1911, the Bradleys, another prominent New York family, also donated acreage to the state of New York. As the Harrimans had, the Bradleys stipulated that their land donation was to be used solely for park purposes. Given by the heirs of Stephen Rowe Bradley, the parcel consisted of a 212-acre plot located on the summit of South Mountain. This land provided a majestic view of the Tappan Zee Bay and the Hudson Valley and was incorporated into the interstate park lands early on, eventually becoming part of what is now known as Blauvelt State Park. At the same time, Dr. James Douglas of New York City donated a scenic parcel located in New Jersey on top of the cliffs at Fort Lee, which offered a magnificent view of the river and New York City, while serving as the southern boundary of the interstate park. By 1912 the commission had also acquired all of the Manhattan Trap Rock properties at Hook Mountain on the Hudson just outside of Nyack, in Rockland County. Other quarry properties that had not been available for purchase were in the process of being condemned, according to the jurisdictional authority granted to the PIPC, as a first step in acquisition proceedings.[19]

By the autumn of 1910, the required stipulations set forth by Mary Harriman had been met, and on October 29 representatives of the PIPC gathered with members of the Harriman family, their friends, and supporters to formally commemorate the family's generous and unprecedented gesture. At the event, eighteen-year-old William Averell Harriman (future governor of New York) made his first public speech and presentation. Following through on the wishes of his father, Harriman presented to George W. Perkins Sr., who accepted on behalf of the PIPC, the deed for 10,000 acres along with checks totaling $1 million (in today's market this sum would be worth approximately $31 million!), made out to the state of New York. At the same time, the state signed over to the PIPC the 700 acres that had originally been designated as the prison site. The agreement stipulated that the state of New York relocate the prison to the eastern shore of the Hudson.[20]

Others present at the festivities that day included Mary Harriman and her daughter Carol Harriman; Mrs. Evelyn Perkins and her daughter, Dorothy Perkins; Mrs. J. Pierpont Morgan and her daughter Mrs. Herbert Satterlee; Mr. and Mrs. Henry Phipps; Dr. Edward Lasell Partridge; former New York governor Benjamin Odell; J. DuPratt White; and William J. McKay, of the Prison Commission.[21]

An article in the *New York Times* recorded the scene: "Nature itself had provided the setting. On three sides of the little plateau (where the ceremony was held) were the hills which form the rugged group clustering about Bear Mountain; just across the Hudson towered Anthony's Nose, jutting out into the river."[22]

Fourteen years and one month later, another momentous ceremony would be held in almost the same location: the opening of the Bear Mountain Bridge.

10

The Chief Engineer

In 1912, a young engineer named William Addams Welch, working for Charles Wellford Leavitt, was selected to handle several projects within the Palisades Interstate Park Commission. While employed by Leavitt, he had designed the Havre de Grace racetrack in Maryland and the boardwalk in Long Beach, Long Island.[1] During the time that Welch was assigned to the PIPC, he caught the attention of George W. Perkins Sr., and Perkins soon drafted him to work directly for the commission as a staff engineer. He proved to be a perfect fit for the agency, and, before long, Perkins realized that Welch was the candidate he had been searching for. He shared Perkins's dream of preserving the wilderness while making it available to the public and had the same desire to teach those persons who visited the park to love and respect it. In the words of Ruby M. Jolliffe, who joined the PIPC some years later as director of group camps (1920–48) and was Welch's colleague, he had a "breadth of understanding and a wide knowledge of the principles of engineering" and "an innate love of nature in all its forms, which evolved into the knowledge and understanding of a great naturalist."[2] Welch and Perkins easily developed a close relationship. In 1914, Welch was named general manager and chief engineer of the PIPC, and together he and Perkins worked to bring their shared ideals to reality. As land was acquired, the commission's task was to incorporate it into the park system in the most cohesive manner possible. Welch and Perkins collaborated to provide guidance and direction, creating the recreational paradise they all envisioned. The Bear Mountain State Park was opened to the public in 1913, following the integration of the Harriman donation with other acquisitions of that time, and after some necessary modifications were made, the gates were opened and attendance was off to a great start. The Bear Mountain State Park was the first of many beautiful retreats that today constitute the Palisades Interstate Park system.

Major William Addams Welch was born on August 20, 1868, in Cynthiana, Kentucky, to Captain Ashbel Welch and Priscilla Addams Welch. Both of his parents had deep roots in the history of America. The first John Welch landed in Plymouth, Massachusetts, sometime after the pilgrims of the *Mayflower* established the first American colony there. William's mother, Priscilla Addams, could trace her ancestors back to John Adams, signer of the Declaration of Independence and second president of the United States. The extra *d* in her last name was the result of certain family members disagreeing with another ancestor, John Quincy Adams, the sixth president of the United States and son of John. The southern arm of the family conflicted with the sixth president's politics and so altered the family name to distance themselves from his ideas.[3]

During the Civil War, Captain Ashbel Welch fought for the Confederacy as a member of Morgan's Raiders, a cavalry unit that had the distinction of conducting the northernmost raids on Union forces, in Indiana, Ohio, Kentucky, and West Virginia. After the war, the family moved to Colorado Springs, Colorado. William grew up in Colorado Springs and attended high school there before moving on to Colorado College, from which he received his degree in civil engineering in 1882. He went on to get his master's of engineering in 1886, at the University of Virginia. Immediately following his graduation, he seized the opportunity to take a grand tour around the world before beginning his career. On November 9, 1899, he enlisted in the US Army, and for the next six years, he worked as an engineer for the US government in Alaska, where he was involved in selecting sites for US forts. While in Nome, he participated in the assembly of one of the first prefabricated iron steamships in the country.[4]

While not too much of his personal story is available through standard historical sources, the information that is available paints the picture of a smart, strong, capable young man immersed in his career. Comprehensive Welch family lore offers information that fills in some gaps in his personal story, such as where and how he met his wife, Camille. According to William's granddaughter, he attended an opera one night and became smitten with the soprano. After the performance, it is said that he ventured backstage and introduced himself, and the rest is history.[5] Camille Beall was an attractive woman seven years younger than Welch. Born in Minnesota on April 24, 1875, she was the daughter of E. W. Beall and Camille Beall. She and William fell in love and married in 1904,[6] and a daughter, Jessie, was born the following year. By this time, William, still serving in the US Army, had been relocated from Alaska to Mexico and South America, and he continued to serve while Camille stayed at home, in New York, with young Jessie. It was a challenging time. In 1906, William contracted yellow fever while working in South America and was forced to return home to recuperate. However, he returned to South America the following year to work on hydroelectric projects and land reclamation. According to the ship manifest for the SS *Obidense*, which sailed from Para, Brazil, on March 22, 1908, William A. Welch, a civil engineer, was listed as a passenger who arrived in New York City on March 26, 1908.[7] This itinerary comports with a family narrative indicating that at about this time, William and Camille agreed that he needed to return to civilian life and come home to New York, where his young family was waiting for him. Census records of 1910 show that William and Camille and their daughter, Jessie Welch, now five years old, were living with Camille's family at 56 Wardell Street, Queens, New York. Included in the household were Camille's father, E. W. Beall, who was now sixty-three years old; her brother, Ople; her sisters, Dorothy Beall and Jessie Beall Cook; as well as her brother-in-law, Chester A. Cook.[8]

It was a busy time for Welch, and not just in his professional life. In 1914, at around the same time he became general manager of the PIPC, he and Camille welcomed their second child, William A. Welch Jr. Welch settled into his home and family life as well as his new position within the PIPC. The relationship between him and George W. Perkins Sr. continued to grow.

Perhaps one of the best demonstrations of Perkins and Welch's collaboration is the Bear Mountain Inn. At the annual meeting of the PIPC in September 1914, a discussion arose regarding building a restaurant that would serve as a symbol of the park and a destination for those individuals who traveled from the city to the pastoral environment. It was decided that this restaurant would be modeled after the Old Faithful Inn in Yellowstone National Park, which combined rustic charm with understated elegance. The Bear Mountain Inn would be designed by the architectural firm Tooker & Marsh.[9] The chosen site was on the old Fort Clinton battlefield, next to Hessian Lake within the northern boundary of the recently opened Bear Mountain State Park. William A. Welch, the PIPC's new chief engineer, would oversee the project. Estimates came in at nearly $100,000, but Perkins and Welch were confident they could deliver the project well below that figure if they could proceed on their own via privately raised funds, without concern for the typical red tape that accompanied government contracts.

Construction began in 1914, and Perkins was able to raise the money through private gift donations, predominantly from the commissioners themselves, with Perkins's gift being the most generous. Welch used PIPC labor crews for the construction and completed the fifty-thousand-square-foot restaurant by early 1915; the project progressed rapidly because they did not adhere to normal building procedures, which would have created a slower pace. Of course, because of the dismissal of normal protocols, there was controversy, and the men were criticized by those persons who supported the state-required process, which was a considerable group. The majority of the building material was garnered from the site. Rubble stone and American chestnut timber that otherwise would have been discarded because of an infestation of blight were utilized in the construction, thus keeping costs down and delivery times tight. The Bear Mountain Inn formally opened on June 3, 1915, beginning a long and illustrious history of catering to park patrons. Besides the many average citizens who have visited the inn, there have been many famous personalities—whose stories have been told over and over during the years that followed, further enhancing the park's history. During the war years in the 1940s, sports teams such as the Brooklyn Dodgers, New York Giants, and Green Bay Packers used the inn as a practice retreat, and many college teams stayed at the inn when they came to play the Army team at West Point. Besides sports figures, the inn featured big-band musicians of the era like Tommy Dorsey and Harry James. And the story is told that Kate Smith sat at a table at the inn while she composed the song "When the Moon Comes over the Mountain."[10]

By 1915 the Palisades Interstate Park Commission had established itself as a permanent part of the valley. Managing the parks was challenging, but the team of commissioners, with Perkins, Welch, and White in the lead, was finding its way, with each participant contributing valuable input in an area relevant to his background and experience. Along with the typical issues encountered, there came a range of unexpected matters that had to be addressed but had no precedent as far as Welch, Perkins, White, or the other PIPC managers were aware. For instance, there was the matter of an unmarried, unescorted single woman applying for a camping permit, for one. Another issue arose involving a married woman who applied for a permit for herself and a group of Campfire Girls, whom she would chaperone, to camp on the park grounds. These issues were delicate at the time, and the commissioners were unsure how to handle them. They decided to establish a "women's only" area for such patrons. Regulations allowed wives to accompany their husbands for camping, but they would have to separate on arrival, with the married women also required to stay in the "women's only" section. They discouraged single women from arriving on their own.[11]

Statistics reported by Welch indicate that 2,369 young women stayed overnight in 1916, a surprisingly large number. In addition to working with nationally known children's organizations, such as the Campfire Girls and the Boy Scouts, new programs were set up involving children of low-income families and disabled children. One such program, established by the New York Association for Improving the Condition of the Poor, sought to improve the nutrition of low-income children in their care. During their stay at the park, the organization would track the results of giving these children healthier meals. Reports indicated that the children's health typically improved, and they gained an average of three and a half pounds each.[12]

The camping programs that had evolved within the PIPC early on by J. DuPratt White continued to be developed and expanded. The earlier permit requirement put in place by White was modified to eventually include a fee of one dollar per week of camping. The PIPC dispatched patrols throughout the parks regularly to ensure that campfires were properly attended, trash was disposed of, and all campers (and other park users) enjoyed their experience, stayed safe, and followed the regulations put in place by the PIPC staff.

Improving transportation to the parks was another important component of Welch and Perkins's plan, and by 1914 almost 115,000 park patrons were arriving daily by day-liner boats. As many as 15,500 came by special charter; ferry patrons accounted for roughly 199,973 visitors by 1915.[13] Many of the day-liner boats were privately operated, but in 1916 the PIPC contracted for a new vessel to be owned and run by the PIPC. The Mathis Yacht Building Company of Camden, New Jersey, would build the *Palisades*, a 120-foot vessel, to transport poor mothers and their children from New York City to the park, free of charge; correspondence between Perkins and Welch revealed that the *Palisades* was the first steamship to be diesel powered. Eventually, the PIPC purchased two additional day liners, the *Clermont* and the *Onteora*, an undertaking that would have unforeseen consequences before long. Under the guidance of Welch, the boats were integrated into the park's evolving operations: schedules were drawn up, maintenance programs were developed, and a myriad of logistical details were addressed as the boats were put into service.

By 1916, the park was recording close to 500,000 visitors annually, while other parks throughout the country were welcoming crowds on an annual basis numbering only in the hundreds.[14] Welch's passion for the outdoors and his wish to provide a natural experience to young and old alike were having an effect.

Stephen Tyng Mather, director of the newly created National Park Service in Washington, DC, took notice of the PIPC's popularity, and in 1917 he invited Welch to speak at the first National Parks Conference. Welch's appearance was treated as a celebrity event; he was introduced as a man "doing really evolutionary and revolutionary" work. In his typical humble style, the major demurred, identifying himself as a "little kitten" among "lions." His speech focused on the many problems the PIPC had encountered and the solutions he and his park staff had devised to address them.[15] Welch and Henry Fairfield Osborne, trustee of the American Museum of Natural History, who accompanied Welch to the conference, extended an invitation to the supervisors of the national parks to visit the New York–New Jersey area and see the PIPC system for themselves. After the conference, Welch's fame spread among park advocates throughout the country, and as his reputation grew, respect for his opinions and expertise increased as well. Volumes of correspondence identifying problems and requesting information and solutions from "the expert" arrived at the PIPC offices daily. His involvement in park management at a time when the idea of national parks was just coming into vogue allowed him to hone his apt skills in this area, developing a philosophy and creating standards of oversight that would become a prototype for park managers worldwide. During his tenure he traveled extensively, providing support and advice to other park managers in the United States and Europe as they struggled to master this new field of study. As a representative of the Palisades Interstate Park Commission, he lectured and assisted others in creating well-run recreational parks. It's little wonder he became known as the "father of the state-park movement."

In line with his love for the park, Welch introduced the idea of a walking path traversing the Delaware Water Gap, through New Jersey and New York, and into New England. The rising interest among the public in walking through nature was reported in articles in the local newspapers, confirming the formation of trail groups such as the Green Mountain Club, Appalachian Mountain Club, Tramp and Trail Club, Fresh Air Club, and many others. In September 1920 several prominent members of the community called a meeting at the Waldorf Hotel in New York to discuss the formation of a "League of Walkers." Welch had a prominent role in the effort and encouraged a number of the already existing hiking groups to meet with the league and continue discussions on October 5. "Hiker's joy is almost here—the mellow month of October, the days of the reddening leaf! Already, with September's crisp nights and mornings, the walker begins to feel the urge to get out into the open," wrote Meade C. Dobson in a September 1920 issue of the *New York Evening Post*.[16]

Welch took the lead in the effort to improve and expand the PIPC trail system within the park in an effort to make the park more usable to pedestrians and tramping organizations. The formation of the Trail Conference worked to increase the number of trails in the park system, but at the same time, it provided an added benefit to Welch by attracting volunteers to the park who kept a watchful eye on the trail infrastructure while ensuring that park regulations were being followed. On April 25, 1922, the organization was formally reorganized, and the "New York–New Jersey Trail Conference" was born, with Major William Welch as chairman. Within a year the conference published the first *New York Walk Book*, a guide to the trails on the New York side of the park. Ray Torrey, a journalist with the *New York Evening Post*, and Frank Place Jr., a conference member, collaborated on its publication, enlisting Robert L. Dickinson, a pencil artist, to provide detailed maps of the trails and facilities a hiker would encounter.[17]

A groundbreaking article appeared in the *Journal of the American Institute of Architects*, in 1921, written by Benton MacKaye of Shirley, Massachusetts. MacKaye proposed the creation of a hiking trail that would extend from Georgia to Maine, a distance of more than two thousand miles and inclusive of fourteen states. His philosophy was that citizens should learn how to survive and live within nature. Doing so would allow them to put life into the proper perspective, enabling them to appreciate the benefits that nature affords and the advantages of taking leisure time in the outdoors. He also advocated for the Boy Scouts and Girl Scouts, declaring that "the scouting life . . . is vital in any real protection of home and country."[18]

Welch developed a close partnership with Ray Torrey, as the two focused on building trails throughout Harriman State Park. They joined with Benton MacKaye to further the vision of a trail spanning fourteen states and two thousand miles, providing a boost of energy to the effort. The first shovels for the multistate trail project were put into the ground at Bear Mountain, in the summer of 1922. The Bear Mountain Trail, the first section of the larger path to come, was completed on October 7, 1923; it was sixteen miles long, beginning in the Hudson Valley, rising from the shore of the Hudson to the summit of Bear Mountain, and from there it descended into Harriman State Park and meandered westward toward the Ramapo River. It was a minimal segment of the entire proposed mileage, to be sure, but it was proof that this challenge was possible—and that trail-loving volunteers could and would create and maintain it! The markers guiding the hikers along their way were designed and created by Welch and installed by volunteers along the trail. The new trail was quickly met with approval from the public. An article penned by William H. Carr, assistant curator in the Department of Education at the American Museum, spoke for many when he expressed appreciation to the numerous volunteer groups that provided general maintenance of the trail. "The influence of their accomplishments will linger as long as there are people who seek the open for rest, recreation, and cultural pursuits," Carr wrote.[19] That small segment of the larger vision would grow slowly as it expanded to the fourteen states MacKaye proposed. In 1968, the US Congress designated what has become the Appalachian National Scenic Trail as a part of the National Park System: 2,158 miles of sweat and accomplishment. The lowest elevation point of the entire Appalachian Trail remains at the shoreline of the Hudson River in the Bear Mountain area. MacKaye's legacy now boasts 2 to 3 million visits annually at various portions of the trail. Among aspiring "thru-hikers"—the term for those individuals completing the entire 2,000-mile trek—as many as 1 in 4 makes it from start to finish.[20]

Welch was dedicated to bringing the park to a level far above the average of what was being offered at the time. He coordinated amenities, including hiking trails, camping, picnicking, and other experiences, and he worked diligently toward that end. While he was involved in hands-on efforts to improve the park, he was also inundated with management issues, and early on, he decided it would be wise to delegate oversight of the park's camping program to Edward F. Brown, who was installed as the first camp manager, under Welch's watchful eye. In the eight years Brown served, he turned the abstract (though haphazard) idea of "community camping" into a systematic program with rules, regulations, and best practices. As a result, the PIPC provided uplifting outdoor experiences to millions of adults and children from the New York and New Jersey areas. The park's popularity grew steadily, with attendance in 1920 reaching 2,166,455 (New York, 1,307,089; New Jersey, 859,366), according to PIPC statistics. The same year, the PIPC had added 331,477 acres to its already generous property holdings,[21] and revenue from the group camps brought a net profit of more than $12,000 for the agency, which was immediately earmarked for reinvestment into the camps.

As the programs began to show a profit, Brown saw an opportunity to request a raise in pay, but the request was denied, so he resigned. A replacement was sought immediately, and within a short period, Ruby M. Jolliffe was offered the position. "Jolliffe," as she was affectionately called, had served as director of camps for the New York City YWCA for the previous eight years and was thus well known by many PIPC commissioners. According to Jack Focht, director of Trailside Museums, she had a rare combination of "dignity and devil-may-care."[22] While she often joined the campers in various fun activities, she had a presence that elicited respect and restraint. Her regular inspections of each park system's camp incited a flurry of activity among the staff charged with ensuring all was in order. Jolliffe's stamp of approval was a highly sought-after prize. With Welch at the helm and Jolliffe assisting, the camping programs of the PIPC flourished.

The year 1914 marked the beginning of World War I, but it was not until 1917 that the United States entered the conflict. Many American boys, both young and old, enlisted to fight for their country, one of whom was George W. Perkins Jr., son of the president of the PIPC, who was twenty-two years old and had just graduated from Princeton University. Welch was already fifty years old when his country called him back into service on May 18, 1918. He entered the US Army once again, this time with the rank of captain, ASSC. Assigned to the Aviation Section of the Signal Corps, his mission was to build airplanes from spruce timber harvested in the Pacific Northwest. By August 2, 1918, he was promoted to major. This promotion did not go unnoticed by his colleague George Perkins, who was appropriately impressed with his friend's accomplishments, and he began to call him "Major" as a nickname, which would stick with him for the rest of his life.[23] Welch served until February 28, 1919, when he was honorably discharged. The war was over, and he returned to his duties at the PIPC.

11
Like Father, Like Son

After the war, George Jr. returned home and worked in several capacities before finding his way to the Palisades Interstate Park Commission. Some years passed before he was appointed commissioner to the PIPC and eventually elected its president, following in his father's footsteps. Besides his involvement in the local community, George Jr. also contributed to his country on a higher level, first by serving in the army during World War I and World War II and later by holding several positions within the federal government, the most significant being his involvement with the formation of NATO.

Early in life, he was exposed to the importance of service to his community and his country. Having his father as a role model was a strong influence on him. George Sr. enjoyed his children and maintained a close, affectionate bond with both his son and his daughter. Reading through his papers, there were many notes written to friends in which he expressed pure delight as he recounted stories about his children and their antics. As they grew, so did his pride in them. On May 2, 1916, George Jr.'s twenty-first birthday, his father wrote: "It is mighty seldom that a father can say to his boy on the day he comes of age that he has never caused his father an hour of serious concern or anxiety. . . . As we pursue our joint journey my fondest desire is that we may, as men, be as good pals as we have been in the past."[1]

As a child, George Jr. attended the Hill School, a preparatory school in Pottstown, Pennsylvania. After graduating in 1913, he entered Princeton University[2] and became immersed in college life.

While on summer break during his college years, he worked as a cub reporter for a New York newspaper. He also worked for the mayor of

New York City John Purroy Mitchel, known at the time as a widespread reformer within city government. Because George's father was an active member of the Republican Party, and, in fact, one of the founding members of the Progressive movement within that party, George had access to political circles at a young age. Memorabilia found among George Jr.'s papers indicate that he served as a page at the Republican National Convention in 1908. At some point, he was also appointed assistant sergeant at arms and served in that capacity for at least one day during the convention of 1912, held in Chicago. A ticket stub for the National Progressive Convention held in August 1912, in Chicago, confirms that he also attended that event on August 5 of that year.[3]

During his senior year at Princeton, George Jr. was introduced to Katherine Trowbridge, daughter of Professor August Trowbridge, head of the Department of Physics at the university. He was quickly smitten, and not too long afterward they announced their engagement. One week after graduation, on June 19, 1917, they were married at the Trinity Episcopal Church in Princeton. After the ceremony, a reception was held at the home of the bride's parents. Immediately following the festivities, the couple left on an extended wedding trip and upon their return settled in New York City.[4] With the Great War raging, the timing of the nuptials was key; on September 11, George Jr. was drafted into the US Army Expeditionary Force as a private. No men were spared from the current events, not even the upper class. Roland Harriman, who was the same age as George, had enlisted after graduating from Yale and secured an officer's commission.

By early 1918, George, now a sergeant in the 304th Field Artillery Unit, was selected from among several candidates to attend a divisional training school for officers at Camp Upton in Yaphank, Long Island. His affiliation with the YMCA provided the young soldier with an opportunity for exemption. It was reported that friends of his went to the local draft board to request an exemption for him. When he was informed, he told the chairman to "throw the exemption claim into the waste basket," as he would be reporting to Yaphank.[5] Shortly after completing training, he was commissioned as a second lieutenant[6] and shipped overseas with the Seventy-Seventh Infantry Division, composed largely of Long Islanders and New Yorkers who would soon participate in the critical battle of the

Argonne Forest in France. While in France, George was transferred to the First Division, and he marched into Germany with them shortly before the Armistice was signed. At about the same time, his father-in-law, Professor Trowbridge, also a member of the armed forces, was shipped overseas, leaving young George's wife, Katherine, now expecting their first child, at home with her mother and her in-laws, George and Evelyn Perkins.

George Sr. was extremely fond of his daughter-in-law and often invited her to Glyndor for visits. On one occasion in 1918, George Sr. sent an invitation to Katherine to come and stay for a time, but she responded that she would be unable to accept as she was not feeling well and suspected she was coming down with a cold. As her cold progressed to include a fever, her mother contacted George Sr. for assistance in locating a nurse; wartime shortages, compounded by an outbreak of influenza, had resulted in a scarcity of nurses in Princeton. George Sr. was able to locate a nurse and a doctor and personally delivered them to his daughter-in-law in Princeton.[7] Despite the intense efforts expended by the medical team, her condition worsened. She died on October 7, 1918, shortly before the war ended,[8] with her mother and father-in-law at her side. George Sr. was devastated. "We have lost the fight, and it is so hard to accept the loss," he cabled George Jr. in France.[9]

He was consumed with the desire to cross the Atlantic himself and comfort his son. Within two months of his daughter-in-law's death, he was on a ship sailing for France to do just that. The Armistice ending the war was signed on November 11, 1918, and while George Sr. arrived in France primarily for his son, once there, he became involved in the monumental task taken on by the United States for the rebuilding of France and Germany following the devastation of the war. He was invited to work with the United War Work Council through his affiliation with the YMCA, a primary member of the council. George Sr.'s father had been instrumental in forming one of the first chapters of the YMCA in Buffalo, New York, in 1852, and George Sr. continued the family association, becoming a national director of the organization in 1901. The stress of losing his daughter-in-law, combined with the heavy workload involved with the War Council, soon took its toll, however. Within a month of arriving in Europe, George Sr. took ill and was confined to the American hospital at

Neuilly, France, for several weeks.[10] Letters between William Welch and Perkins during this period highlight the concerns of both men for their country, but, at the same time, acknowledge their continuing dedication to the commission. Welch, who returned to New York before Perkins, explained that he was working diligently to settle matters at the PIPC and get things back on keel after his absence.

Perkins never fully recuperated from the pneumonia that had sent him to the hospital in France. Not only was he mourning the loss of his daughter-in-law and the impact the tragedy had on his son, but while he remained overseas, he was involved in the Red Triangle program, the name for the YMCA agenda that served American soldiers who could not immediately return home after the war. Certain troops were held in Europe, maintaining a US presence until all conditions were met by Germany and the peace accords were finalized. George Sr. finally returned home in 1919, but he continued to struggle with his health for more than a year after his return, attempting to regain his strength. After the war ended in 1918 as the majority of American soldiers returned home from Europe to their normal lives, Welch and Perkins continued their partnership. Even with so much else consuming his attention, Perkins's dedication to the PIPC was never far from his mind, and from his hospital bed, he continued to communicate with Welch and assist him with ideas and possible solutions to management issues.[11]

George Jr. carried on the family involvement in the YMCA, and rather than immediately returning home once his duties were complete, he too signed up to work with his father in the Y's Red Triangle program, extending his service overseas and delaying his return to New York until 1919.[12] Once he returned to the United States, he threw himself into service to his country and community to assuage his grief over losing his wife and child. He continued his affiliation with the New York City YMCA and advocated for the creation of a committee made up of Princeton graduates to study social, civic, and religious issues within the community. In a press release designed to help Perkins entice candidates for his committee, he was quoted as saying, "We need a great many college men and, frankly, a lot of them need the very thing we offer—contact with the big problems of life. If they are to be the leaders of tomorrow, they need to know the thought of the people today, since that is to eventuate into the actions of tomorrow."[13] Besides his work with the YMCA, he served as a member of the newly formed American Committee for Aid in the Republic of Poland (a cause supported by several prominent New Yorkers, among them Lieutenant Colonel Theodore Roosevelt, the son of the former president);[14] he also served as executive secretary of the Princeton Endowment Fund Committee and as a member of the executive committee of the Boy Scouts of America.

While George Jr. was busy working to educate and redirect his generation to develop competent world leaders, his father, having returned home as well, was trying to follow his doctor's orders and avoid overexertion. The pneumonia he had contracted in Europe left him tired and frail, and he had developed a heart condition. Regardless, Perkins Sr. was not one to sit idle for any length of time. He continued his fund-raising efforts and property acquisitions within the PIPC. In February 1920 he experienced another physical setback when an intestinal condition flared up while he was wintering in Florida.[15] His health continued to deteriorate, and after returning to New York, in April 1920, he was sent to the Catskills on doctor's orders for a complete rest. Shortly after returning home from this respite, he suffered what was thought to be a nervous breakdown, presumably the result of complete exhaustion brought on by his unyielding work schedule and general life overload. He was checked into a sanitarium in Connecticut to fight the condition but to no avail. He died on June 18, 1920, at the age of fifty-eight. His doctors gave the primary cause of death as acute encephalitis, or inflammation of the brain, and the secondary cause as chronic myocarditis, a cardiac condition they believed worsened because of influenza and pneumonia that he contracted in France.[16]

George W. Perkins Sr. had risen far above his father's expectations for him, and he became one of the most prominent men of the early twentieth century, contributing his all in whatever he took on. According to an obituary written at his passing, "At a time of widespread unrest, when passions and prejudices are lashed to high heat by violent and bitter agitation when the spirit of conciliation and compromise is needed for readjustment, the nation can ill afford to lose such a spirit as that of George W. Perkins."[17]

George Jr. was named executor of his father's estate and assumed the role of managing the family's affairs from that point forward.

After the death of George W. Perkins Sr., William Welch was quoted as saying his world "turned upside down."[18] The absence of his good friend and colleague made managing the PIPC considerably more difficult, though he continued the work they had both committed to with spirit and dedication. In addition to the myriad responsibilities Welch now shouldered, he spearheaded a campaign to erect several memorials to the man who, in his mind and many others, had been the driving force and inspiration for the commission from the beginning. The memorials were completed in 1932. Perkins Memorial Drive provided—and still provides—access to the top of Bear Mountain from which vantage point a magnificent view of the entire park is available. In cooperation with the states of New York and New Jersey, Welch and the PIPC built a tower on the summit as a monument to Perkins, "whose broad vision and tireless energy made possible the preservation of the Palisades and the establishment of this great playground for humanity," as a placard on the site explains.

It was 1921 when George Jr. received his master's degree from Columbia University in New York and then headed to Washington, DC, for his first civilian assignment since the war; he was appointed executive secretary to Postmaster General Will Hayes.[19] In Washington, George met Caroline (Linn) Merck, and soon afterward they announced their engagement. George and Linn were married at the Grace Church on Lower Broadway, in New York City, on December 17, 1921, with their families in attendance,[20] and afterward celebrated with a small reception at the home of the bride's brother, George W. Merck, in Manhattan.

Caroline (Linn) Merck Perkins was the daughter of Mr. and Mrs. George Merck, of Merck and Company, a successful family business producing pharmaceuticals and chemicals. The company originated in the 1600s in Germany when Friedrich Jacob Merck purchased a drugstore in his hometown of Darmstadt. The store evolved into a drug-manufacturing facility when a process to extract morphine from opium was discovered. By 1800 the company had grown considerably. In 1891, George (Georg) Merck, the grandson of the founder, along with an associate of his immigrated to America, bringing with them the blessing of the family and

$200,000 to establish Merck and Co. in Manhattan, as a subsidiary of the German parent company. Over the years, Merck and Co. experienced many ups and downs, and, eventually, the American branch became a separate corporation with its own identity and was awarded the sole right to use the Merck name.

After their wedding, George and Linn settled in Washington, DC, making their home at George's residence until his tenure at the Postmaster's Office was over and they returned to New York City. Once back in New York, George took a position with G. Amsinck and Company, Inc., as an import-export broker. He was also named assistant treasurer of the Republican State Committee of New York and served on the Executive Committee of the Young Republican Club.[21] On March 14, 1922, George W. Perkins Jr. was nominated by Governor Nathan Miller of New York along with George T. Smith as commissioners of the Palisades Interstate Park Commission, the organization so treasured by George's father.[22] The men were named to the PIPC as a result of the deaths of two sitting commissioners: John J. Voorhees (originally appointed in 1915) and Otis H. Cutler (who had been appointed in 1921 to fill the spot vacated by G. W. Perkins Sr.'s death). Voorhees and Cutler died in March 1922, within a week of each other, and the new appointments followed quickly. With his nomination and acceptance, George Jr. acknowledged his intention of forging a path of his own, although his journey could not help but be traveled in his father's shadow. As the first president of the PIPC, the senior Perkins had created the commission as it existed, and he had been the driving force behind its successful development. Continuing such a legacy would be a tall order, indeed.

12
We Must Not Forget

The twentieth century had begun as the Palisades Interstate Park Commission first took its place, and while the world experienced the "Great War" life continued in relative normalcy within the confines of the Hudson River valley. The PIPC developed and preserved the land, and the New Jersey State Federation of Women's Clubs continued to support their work. Together they worked diligently to achieve goals only imagined at the onset. The women worked side by side with the PIPC and stood ready to assist whenever they were called upon, whether through the NJSFWC, the Englewood Women's Club, or the newly formed League for the Preservation of the Palisades.

As they worked in unison with the commissioners developing the commission's legacy, the women never forgot about their request to commemorate those individuals whose original efforts were key in the commission's formation. The commissioners had agreed to the request for a memorial, but within the agency, there was no active support for it.

Nonetheless, the women continued to press the issue, as correspondence makes clear. Letters dated March through May 1915 between J. Du-Pratt White and S. Elizabeth Demarest (secretary of the NJSFWC) reveal Demarest's fears that if too much time passed without action, the women and their memorial would indeed be forgotten. She voiced her concern to White that once he was no longer at the commission, the "powers that which shall then have arisen will not know us either and will forget even for what those few thousands (of dollars) are for."[1] The back-and-forth continued until a general discussion of monument style was initiated. White recommended that the design be "rugged" and thus more difficult to vandalize, as the parcel of land for the women's park was remote and isolated. Even getting to this point was not easy—a small but not insignificant step; however, a new problem arose. Costs were rising, and the expense of building a monument remained consistently just out of reach. Still, White continued to support the women and the project, at least according to his letters.

By 1919, as the effects of the Great War settled down, J. DuPratt White attempted to resurrect the matter of the memorial once again with the other commissioners. He enlisted the aid of Henry G. Emery, an architect friend of his, to provide some informal sketches of possible designs for the women to review. However, there is no evidence that any sketches were ever submitted, any selections were ever made, or any progress of any kind resulted from his efforts. Several years later, on March 14, 1921, NJSFWC president Ida W. Dawson wrote to White referencing the original March 1908 agreement, asking, her frustration palpable, "Can you inform me as to what has been done about it [the memorial] in the last thirteen years . . . ?"[2] White responded the following day, explaining that the commission had been holding the money forwarded by the NJSFWC at the time of the agreement, but since "the amount was not nearly sufficient, nothing was done."[3] On March 20 Mrs. Dawson responded to White's note and explained that "the Federation has other matters on hand now, so cannot collect further funds for the Palisade memorial. While I realize the money on hand is not enough for a pretentious memorial, yet could you not use it for something that would serve as a marker . . . ? Perhaps you could make a cement seat and railing and an inscription in tablet form." The most important thing was to finally act after so much delay, she argued. "It would be advisable to do something now, while all are living who are interested. Do you not think so?" she asked.[4] White agreed and promised to "give the matter some thought." Correspondence discovered suggests that there was no additional movement on the project until many years later when, in June 1928, William Welch sketched three potential monuments and sent them to Lydia S. Osborne, then chairman of the federation's Memorial Committee. This correspondence and exchange played out against the many changes and issues of the early twentieth century. The growth of the railroad was a major development, as was the introduction of the automobile, and the effects of these two issues played an important role in how the world was evolving.

At the end of the nineteenth century, the Hudson River had five railroad bridges crossing its waters, most positioned close to Albany, although one was as far south as Poughkeepsie.[5]

The need for a bridge closer to Manhattan was recognized and discussed, but identifying the right location was a challenge. There were many locations identified and subsequently investigated, but because the cost to construct a bridge at most of these locations was found to be exorbitant, each in turn was abandoned and other locations were researched. The point between Anthony's Nose on the east shore and Fort Clinton–Bear Mountain on the west shore was considered an ideal location, as it was the narrowest section of the river below Albany and would dictate more affordable construction costs. There were historical connections to this location as well. The site selected was the very place where, in 1777, colonial forces installed the first of two chains across the Hudson in the early stages of the Revolutionary War. The chain was positioned as a deterrent to prevent the British navy from sending their frigates north along the Hudson River. At the west end of the proposed bridge was the location of Fort Clinton, a colonial military position that was subsequently destroyed by the British in 1778.

Around 1920, the popularity of motorcars surged, and an enterprising bridge builder, Frederick Tench, decided that a vehicular crossing was becoming a necessity. Tench identified that the most likely location for this crossing was the Fort Clinton–Anthony's Nose site. As mentioned, history testifies that he was not the first to suggest a crossing at this location, but he, ultimately, became the man who accomplished that feat, in collaboration with Roland Harriman.

At the turn of the twentieth century, Roland's father, E. H. Harriman, although busy building his railroad empire, had been aware of the changing landscape across the nation. He recognized the evolving need for river crossings, but there was not yet any real urgency to justify the expense and considerable amount of work involved in building one.[6] A decade after his passing, however, the urgency was evident and could no longer be ignored. The growth of the railroads and the invention of the motorcar, also known as the "horseless carriage," brought about a whole new list of reasons to warrant the need for a Hudson River crossing south of Albany.

By 1914 the price of a Model T had fallen to $490, and, in 1919 General Motors and DuPont introduced manufacturer-backed financing options with the creation of the General Motors Acceptance Corporation.[7] The combination of the decrease in automobile purchase prices and the introduction of consumer credit resulted in a dramatically enlarged consumer pool for motorcars in the United States. Increased ownership brought an increased number of cars onto the roads, and by 1922 there were approximately ten million automobiles registered in the United States. A new vocabulary was developed to describe this revolution in transportation, including a term that strikes horror in the mind of the average motorist still today: *traffic jam!*

A report appeared in the May 1923 issue of the *Scientific American*, written by Major William A. Welch, general manager of the Palisades Interstate Park Commission. Major Welch cited that up to that point, all cross-river traffic south of Albany had been by ferry. The ferry companies had been consistently improving service, adding new ferries and expanding their schedules to keep up with the rapidly increasing demand, a result of the growing trend of motorcar ownership as well as the use of motor trucks for commercial deliveries—in short, more vehicles on the road. Although Welch credited the ferry companies for their valiant efforts to keep up, he concluded that try as they may, they were struggling to accommodate the customers on weekdays, let alone the business on Saturdays, Sundays, and holidays.

Welch noted, "It is a matter of common experience for motorists to wait an hour or more for a ferry; many cars are subject to a wait of 3 or 4 hours at various times of congestion; and on Decoration Day, 1921, some cars had to wait as long as ten hours before they could cross on the Dyckman Street Ferry."[8]

Indeed, according to that same article authored by Welch, "on several Sundays and holidays" in 1921, "as many as 5,500 motor cars stopped at Bear Mountain," and "the total number of cars during the season amounted to over 300,000," the majority of which had to cross the river via ferries. Furthermore, according to Welch, over the previous five or six years, car traffic had "increased annually at the rate of about 40% per year."[9] It is little wonder that the consensus among valley residents and visitors alike was leaning more and more toward a bridge being built somewhere south of Albany to carry the automobile traffic, thus alleviating the congestion.

Regional developments exacerbated the problem. By the early 1900s, New York City was considered the most important city in the country; as a result, the city and the area surrounding it had become extremely congested and chaotic. The port was a beacon of trade, finance, manufacturing, and culture, yet there was no river crossing from neighboring New Jersey to the city of New York. Almost all the major railroads had multiple trains arriving daily at New Jersey terminals, and many railroad passengers had New York as their destination. Travelers were thus forced to disembark trains and board ferries to complete the final leg of their journey. The trip across the Hudson, while short, was fraught with scheduling delays owing to the volume of trains on a given day or—more concerning—foul weather that created dangerous conditions at worst and uncomfortable travel at best. Goods being transported also needed to be loaded and unloaded, a costly and time-consuming process. As inefficient as the transportation system was, it was taken in stride. There was no alternative. In the late 1800s, there had been talk of building a railroad crossing carrying six sets of tracks and terminating at a location near Twenty-Third Street in Manhattan. A design had even been completed by well-known engineer Gustav Lindenthal, but it was as far as things got.[10]

For years, dissatisfaction with the status quo prompted discussions on alternatives that might be available to improve the region's transit situation. New York Harbor was more than a mile wide, with harsh currents and extreme depths, conditions that did not lend themselves to building a bridge. The (very high) estimated cost of building at that site stymied the project. A bridge connecting to Weehawken, New Jersey, was contemplated, but again the cost ensured that it never got off the ground.[11]

The location spanning the river from Anthony's Nose to Bear Mountain–Fort Clinton, approximately three-quarters of a mile wide, was the narrowest crossing, and it was located only forty miles north of Manhattan. Additionally, it had several other advantages. Besides its smaller footprint, earlier engineering studies had determined that the proposed approaches on each side of the river had natural footings of solid granite deposits that could be utilized for tower foundations.[12] This advantage would lend itself to building a bridge with minimal engineering challenges and therefore an affordable cost. Here was the bridge Frederick Tench wanted to build. He just needed to find someone to pay for it.

13

A Man with a Plan

Early in 1922, Tench wrote to Mrs. Mary Harriman, widow of E. H. Harriman, outlining his thoughts on building the bridge at Bear Mountain. In a transparent attempt to secure financial backing from the Harriman family, he suggested that the bridge be named the Harriman Hudson River Bridge. Mary consulted with her sons Averell and Roland. All three acknowledged that a real need existed for a vehicle crossing spanning the Hudson, and they agreed that the location Tench suggested would be a good one. However, the Harriman family was not eager to be the bridge's sole financier. E. Roland Harriman responded to Tench on behalf of his mother in a letter dated January 12, 1922. He politely declined to lend the family name to the project, but not before adding that Mrs. Harriman "wishes you to know that she thoroughly endorses the project and is anxious to be kept in touch with it as it progresses."[1] Tench responded to Roland on January 17 with the following: "I appreciate very much indeed the letter that you wrote me on the 12th and am more than pleased to know that Mrs. Harriman should express herself as favorably toward our proposition. I am working as hard as I can on this and feel convinced that we will build the bridge. . . . I shall be glad to keep you advised from time to time. . . . We regret very much that Mrs. Harriman did not feel like using her name in connection with the structure but understand fully why she should not consent."[2]

In the course of their exchanges, a camaraderie developed between the men, and Roland soon found himself helping Tench to raise money and to secure government approvals, even introducing him to several business associates whom Roland considered to be possible investors.

E. Roland Harriman was born on Christmas Eve, December 24, 1895, in New York City, the youngest of six children born to Edward Henry (E. H.) Harriman and Mary Williamson Averell. His life was one of enormous privilege since his father was a wealthy railroad magnate who developed the Union Pacific Railroad, the Southern Pacific Railroad, the Illinois Central Railroad, and many other railroads and business ventures.

Roland loved spending time with his father, and, although the elder Harriman was immersed in the serious work of reorganizing railroads and structuring major business deals, his family remained his priority, and he made it a point to always dine with the family and was sure to include time for the children in his schedule. In *I Reminisce,* Roland's memoir, he recalls frequent visits to his father's study where they would play games together. Henry's influence on the children was great, and his expectations were lofty. While still a young boy, Roland was instructed by Henry on the importance of looking someone in the eye when speaking to them. Henry insisted he do that each time they spoke, telling Roland that he (Henry) had no use for those men who did not look him in the eye.[3]

The elder Harriman's status allowed Roland access to experiences of which few boys his age could boast. In 1899, for example, E. H. Harriman chartered a steamship from Seattle to Alaska for a family vacation. His doctor had recommended that he get away and relax. He soon turned what was to be a family vacation into a research expedition, inviting a group of distinguished scientists, botanists, and geologists affiliated with the Smithsonian Institution to come along with Mary and the children for a three-month all-expense-paid adventure. The research team's discoveries resulted in thirteen new genera and six hundred newly discovered species, all documented for science—enough information to fill twelve scientific volumes. They also took five thousand photographs and generated a sizable collection of paintings and sketches done by several famous artists, who were also guests of the expedition.

Included among the research team was naturalist John Muir. Upon meeting Harriman for the first time, Muir's typical mistrust of businessmen was decidedly challenged. Muir later wrote in a tribute to E. H. Harriman for the Sierra Club, "I soon saw that Mr. Harriman was uncommon. He was taking a trip for rest, and at the same time managing his exploring guests as if we were a grateful soothing essential part of his rest cure, though scientific explorers are not easily managed, and in large mixed lots are rather inflammable and explosive."[4] Muir's most vivid image of Henry was of him keeping trot-step with little Roland while helping the four-year-old drag a toy canoe along the ship's deck with a string.[5] Roland tells us that the main thing that he remembered from the trip was a conversation he witnessed between his father and some of the men, on the topic of his father's wealth and his attitude and priorities regarding it. Henry took little Roland aside later and explained to him, "I have never cared for money except as a power to put to work. I was lucky and my friends and neighbors, observing my luck, brought their money to me to invest, and in this way, I have come to handle large sums. What I most enjoy is the power of creation, getting into partnership with nature and doing good, helping to feed man and beast, and making everybody and everything a little better and happier."[6] Thus was Henry's perspective on life and wealth, according to his son.

Roland was four years old when he was enrolled at Bovee School, described in his memoir as a school "for young brats." He soon transferred to the Pine Lodge School in Lakewood, New Jersey, which was able to provide more one-on-one attention by limiting each class to only six or seven students. In 1907, at the age of eleven, he transferred to Groton School, a private Episcopal college preparatory boarding school in Groton, Massachusetts, and later attended Yale.[7]

Growing up, Roland enjoyed being outdoors with his father. They hunted and fished at Pelican Bay Lodge in a remote location on Klamath Lake in Oregon. It was at this remote lodge that Roland killed his first bear in 1908, a feat often recounted by the young hunter to anyone who would listen.[8] In 1908, Roland's father bought a functioning cattle ranch in Idaho, sight unseen, from the Guggenheim family. While E. H. never did get to visit the land, after his death, Mary and the children used it as a family getaway. They eventually donated the property to the state of Idaho, and on April 1, 1977, the ranch became Harriman State Park of Idaho (not to be confused with Harriman State Park of New York, which was also created on land donated by the family).[9]

By the time he was twelve years old, Roland had the enviable accomplishment of having visited every state in the country, save two, and had done some international travel as well.

Perhaps, if he had lived longer, Henry might have accomplished his dream of creating an around-the-world transportation empire. But, as fate would have it, he did not. At the height of his career, he "controlled 75,000 miles of railroad track, possessed $400 million in stocks and bonds, held another $150 million in cash";[10] in 1909 at his death, his empire was valued in total at between $150 million and $200 million (which in today's market would equate to between $4.9 billion and $6.5 billion). E. H. was a very controversial figure, and his children grew to adulthood hearing many negative stories about their father, as he was vilified by business and political adversaries; nonetheless, based on their family experiences, they idolized him. Their lives were filled with love and privilege because of him. Regardless of how people viewed him, he was a larger-than-life personality, and his shadow would follow them throughout their lives, despite their accomplishments.

The summer of 1915 was one of Roland's most unforgettable. He was twenty years old, and his mother had allowed him to invite a group of friends on the family yacht, the *Sultana*, for a cruise setting sail from Long Island to San Francisco. The route included stops in Havana, Jamaica, the Panama Canal, Acapulco, and San Diego. While most of these destinations were not as developed or luxurious as they are today, it was nonetheless a memorable vacation. Around this same time, Roland's good friend Bob Lovett traveled to Europe on a family vacation. While there he met the Fries family, from New York. Dr. Harold Fries took a liking to Bob and introduced him to his daughter, Gladys. After returning home to the United States, in the autumn of 1915, Bob and Roland found themselves at a party with the Frieses, and Bob introduced Roland to Gladys. This introduction resulted in an engagement some months later, and on April 12, 1917, Roland and Gladys were married.[11] Gladys had graduated from Spence the year before and had a wide circle of friends with whom she enjoyed traveling and social affairs. She worked at the Junior League, and Roland was impressed with her many skills and activities.

With the onset of World War I, Roland applied for a commission with the Army Ordnance Department and was dispatched to inspect several plants that made weapons, ammunition, and equipment. During one of these assignments, he contracted pneumonia and was laid up for some time. After he recuperated enough to return to his duties, he was hit with an attack of influenza—and, soon after, was diagnosed with incipient tuberculosis. As a result, in December 1918, his doctors recommended that he spend a year recuperating in Santa Barbara, California, where the weather was much milder than in the Northeast. In January 1919, he received an honorable discharge from the US Army, and Roland and Gladys traveled to Santa Barbara as the doctors advised. They lived there while Roland convalesced and welcomed their first daughter in February 1919. Not long afterward, they purchased a ranch in the foothills of Santa Barbara and worked the ranch themselves with the help of a local farmer. His health stabilized to a consistent level, and he returned home to New York with his family, although it was several years before he felt confident that he was truly recovered.[12]

While Roland was living in Santa Barbara, his brother Averell started a shipyard in Pennsylvania, not far from Philadelphia. He was motivated to enter the shipbuilding industry after identifying a shortage of merchant ships a few years earlier. As World War I got under way, all available ships were being used for military purposes, leaving an even greater shortage of vessels for merchants to transport their products to market. Frederick Tench and Edward Terry also recognized the need and developed a shipbuilding enterprise in Savannah, Georgia, at that same time.[13] The Harrimans followed suit, and Averell assembled a staff of shipbuilders and opened the Merchant Shipbuilding Corporation in Chester, Pennsylvania. When Roland returned from California, he joined his brother in the financial management of the business. The brothers were successful in this venture throughout the war years and several years beyond.[14] They worked well together, though they were opposite personalities—or perhaps because of it. Sometime in 1922, Averell established an investment firm, W. A. Harriman and Co., with offices at 120 Broadway, New York City. Roland joined him in that venture as well, helping to grow its reputation and success.[15]

When Frederick Tench approached the Harrimans about building the Bear Mountain Bridge, Roland became interested in the project, and, although he encouraged Tench to do the legwork, he was generous with his connections and financial expertise. In a letter written by Roland dated January 27, 1922, he suggested Tench consult with Dr. Edward L. Partridge, an acquaintance of Roland's who knew his way around the New York Legislature. Partridge was a prominent doctor in New York City who had been involved in preserving the Hudson River Highlands and the Palisades.[16] He was also a sitting commissioner of the Palisades Interstate Parks Commission and was often involved in lobbying the New York Legislature on behalf of the PIPC regarding properties and regulations as part of the development of the commission.[17] A part-time resident of Cornwall-on-Hudson, Partridge was also a neighbor and friend of the Harriman family, and his opinions were highly respected within the community. In Roland's correspondence with Tench, he included a letter addressed to Partridge that the bridge builder was instructed to present to Partridge when they met, as a form of introduction:

> This will introduce to you Mr. Frederick Tench of Terry and Tench, who have conceived the idea of building a bridge across the Hudson between Bear Mountain and Anthony's Nose.
> Mr. Tench has interested me very much with his idea and from a letter he has shown me from Mr. Welch, he, too, is very enthusiastic, and likewise my mother. A bill to secure the Charter is ready to be put before the legislature at Albany, but I think that Mr. Tench would like to see you and talk the matter over and, perhaps, enlist your cooperation if you think well of the project. In any event, I would like to receive an indication of your thoughts on the matter.[18]

Tench was extremely grateful for the gesture and expressed his thanks to Roland in a response letter he sent the next day, informing Roland that he had immediately reached out to the doctor and had confirmed an appointment set up for the following Tuesday.[19] Tench met with Partridge as planned. Presumably, he caught the doctor's interest since the charter application for the Bear Mountain Hudson River Bridge Company was submitted to the New York Legislature that March.

14
Baird

Tench was required to assemble a cost analysis of the project to file for state approval. It would include a full set of construction drawings and design details prepared by an accredited engineer. He reached out to Howard Carter Baird, an engineer he had worked with over the years, and asked him to take on the project and create the necessary documents. At this time Baird had his own engineering firm in New York and had worked in the field for almost thirty years.

Howard Carter Baird had begun his career as a draftsman in 1892, working on railway and highway bridges, creating shop drawings and design details at the Phoenix Bridge Company. Young Baird had left his home in Louisville, Kentucky, three years earlier. He settled in Phoenixville, Pennsylvania, and secured a job at the bridge company. Within a short time, he gained recognition as a quick learner and a hard worker and moved through the ranks, steadily improving his position. As time went on, he became an assistant engineer, a role that gave him the responsibility of designing steel buildings, supervising the engineering department, handling bids, and designing erection equipment. His advancement from draftsman to designer and assistant engineer occurred as part of an apprenticeship program typical in the late nineteenth century. Phoenix Bridge Company journals covering 1896 through 1900 list Baird as having prepared design documents for many large projects, among them several elevated railway structures in the New York metropolitan area. They included parts of the Brooklyn and Brighton Beach Railroad, the Long Island Electric Railway, and the Kings County Elevated Railway that connected with the Seaside and Brooklyn Bridge Elevated Railway. He was involved in designing the superstructure that supported the Atlantic City

Boardwalk, the Tioranda Viaduct, the St. Lawrence River Bridge, and more than a dozen other major infrastructure projects—many of which still survive today.[1]

Phoenix Bridge Company was one of the most well-known and highly respected bridge builders in the late 1800s because of the bridges they built and the innovative construction processes they had developed. It was a subsidiary of the Phoenix Steel Corporation, well known for many inventions and innovations in the steel industry. They had a rich history, tracing roots back to 1790 when they were known as the French Creek Nail Works, the first American nail factory, situated near the mouth of French Creek, in the town of the same name in Pennsylvania. In 1813 the company was renamed Phoenix Iron Works and expanded its product line into manufacturing a diversified list of iron products. As railroads became a major industry, the firm expanded into rolling railroad rails. When one of the owners, David Reeves, and his chief engineer, George Walters, invented a "gag press" for straightening rails, a new business was formed. In 1861, John Griffen, superintendent at Phoenix Iron, developed the Griffen gun, and in 1862 Samuel J. Reeves invented and patented the Phoenix Column, a world-famous structural element used in buildings and bridges for many years. The company's reputation grew steadily and was known far and wide. In 1868 Reeves formed Kellogg, Clarke, and Company, a separate entity that would construct buildings and bridges, and finally, in 1884, the company was renamed the Phoenix Bridge Company, which remained in business until 1962, and the village where it was located was eventually known as Phoenixville.[2]

When Baird joined the company in 1892, it was approaching the peak of its success, and he was caught up in the excitement of bridge building and the opportunities available to him at Phoenix Bridge Company.

Baird was born on January 9, 1868, in Louisville, Kentucky, to James P. Baird and Martha Howard Baird. He was the second of four children and the only boy. His sister Mira H. Baird was two years older than him, and he had two younger sisters, Lucy (Louica) and Martha, born in 1871 and 1874, respectively. The family remained in Jefferson County, Kentucky, while the children were growing up, and Howard and his sisters attended the public schools there. Around 1887, nineteen-year-old

Howard moved out, relocating to Phoenixville, Pennsylvania, and securing a position at the Phoenix Bridge Company.[3] Military service documents show Private Howard C. Baird enlisted in the Pennsylvania National Guard on June 16, 1898, at the age of thirty. He was promoted to sergeant in the Pennsylvania National Guard on June 18, 1898, two days after enlistment, and was discharged the following year on June 13, 1899.[4] The Spanish-American War broke out in April 1898, but lasted only four months, ending in August that same year. Whether Baird enlisted because of this war, we could not confirm. Records show that the US government needed trained engineers at that time, and, according to orders issued by the secretary of war between July and August 1898, recruits possessing mechanical skills were needed from every branch of the engineering profession.[5] However, there is no evidence to indicate that Baird ever served in any engineering capacity during his time in the military or that his skill set played a role in his enlistment. William A. Welch, on the other hand, is documented as having served in the US Army working in Alaska and South America as an engineer for the US government at about this same time (1899). After Baird's discharge, he returned to Phoenixville. The US Census of 1900 shows him living at 143 Main Street, in Phoenixville, Ward 2, as one of three boarders renting rooms from the Shoemaker family.[6]

In 1892, when Baird first joined the Phoenix Bridge Company, bridge building was in full bloom across the country. At about this same time, Frederick Tench and Edward Terry were working their way from the Midwest eastward, building bridges across the Mississippi and Missouri Rivers, as well as in the state of Pennsylvania. The Phoenix Bridge Company, as a major player in bridge construction throughout the eastern portion of the country, was very likely involved in at least some of these projects. Phoenix Bridge Company records indicate that while employed at Phoenix, Baird worked on the Harlem River drawbridge in New York.[7] Records also show that Terry and Tench secured a contract for work on that bridge as one of their first projects after coming to New York in 1894. Therefore, one could speculate that Baird was already acquainted with Edward Terry and Frederick Tench before he made his move from Pennsylvania to New York City in 1904, but there is no firm evidence.

When Baird arrived in New York City, he was employed by the Brooklyn Rapid Transit Company for a time, but in 1904 he began work at the firm of Boller & Hodge, an engineering firm with offices in both Brooklyn and Manhattan. He arranged to apprentice under the tutelage of Alfred P. Boller and Henry W. Hodge, both highly regarded engineers.[8] Boller and Hodge had also been involved with the Phoenix Bridge Company earlier in their careers—another possible connection between these men who shared a common profession.

Baird initially worked in a junior capacity at Boller & Hodge, but as he learned more about the technical intricacies of the trade, he was able to move ahead. By 1909 (five years into his eight-year apprenticeship), he assumed his role as a full engineer, designing miles of reinforcement for a double-track elevated railway, awarding contracts for the project, and supervising its erection. After the completion of his eight-year apprenticeship, he was listed as principal assistant engineer at the Manhattan office of Boller & Hodge. In that role, he designed multiple projects, the most important being a major portion of the superstructure for the St. Louis Municipal Bridge, as well as numerous railway spans for the National Railway of Mexico, the Manila Railway, and other foreign lines. In 1912 Baird was elevated to partner, and the firm was renamed Boller, Hodge, and Baird. As a partner in the firm, he designed the superstructure of the East Haddam Bridge over the Connecticut River, containing 460 feet of swing draw span for the highway; the Congress Street Bridge over the Hudson River at Troy; and the St. John's River Bridge between Canada and the United States, all of which featured heavy railway spans.[9] Boller died in 1912, and Baird continued working alongside Henry W. Hodge for the next few years. Baird notes in his dissertation on the Bear Mountain Hudson River Bridge that he and Hodge worked together sometime between 1913 and 1915 on a steel arch double-track railway design at the very site where the Bear Mountain Bridge now stands.[10] However, the project did not move forward. In 1919 Henry Wilson Hodge suffered an embolism and died, leaving Baird as sole proprietor of the firm.[11]

Around 1921, Baird's colleague Frederick Tench was fully immersed in his campaign to secure approvals and financing for a bridge he wanted to build across the Hudson River. H. H. Sherwin, of the Terry & Tench

Co., Inc., had identified a location for the structure: the point at the river between Bear Mountain Park and Anthony's Nose. Tench approached Howard Carter Baird, and Baird agreed to accept the assignment, delivering the design and a full set of construction drawings.[12] Based on Baird's documents, Tench prepared a cost estimate and the balance of the required package and shared it with E. Roland Harriman, convincing him to join the project as well. Tench then applied for the necessary approvals from the state of New York and the federal government, Department of War. The Bear Mountain Hudson River Bridge Company was created according to the state charter, and the company awarded the contract for construction of the bridge to the firm of Terry & Tench Co., Inc.

15

Bear Mountain Hudson River Bridge Company

News spread through the valley about the proposed bridge and generated excitement among the public, reflected in the articles appearing in local Hudson Valley newspapers and trade publications. But there were people who opposed the idea as well. In February 1922, when the first rumblings were brought before the New York Legislature, an article appeared in the *Brooklyn Daily Eagle,* reporting on a new toll bridge being proposed near Peekskill that would cost $5 million, facing stiff opposition. The article noted that "the proposal will probably meet some rough sledding." The state of New York had just adopted a toll-free policy and converted the last toll bridge to a toll-free crossing two years earlier. The new bridge at Bear Mountain was being proposed as a toll bridge, and it was expected to be turned down. It would "connect the Interstate Park with the State Road east of the Hudson and provide a direct motorcar route from Boston to Philadelphia without passing through New York City's congested streets."[1] At the same time, a movement was afoot backing a highway bridge at Poughkeepsie, whose citizens, of course, felt that their town would be a much better place for a crossing. The following month the *Poughkeepsie Eagle News* reported that bills for both bridges were being presented for consideration in the current session of the legislature. The Bear Mountain Bridge proposal was not heard that day owing to changes being recommended before its presentation. Proponents of the Poughkeepsie bridge were elated. They believed that should their bill be heard first, they would enjoy the advantage.[2] Three days later the *Yonkers (NY) Herald Statesman*'s headline read "Bridge to

Bear Mountain from Peekskill Is Passed." The piece explained that the Smith-Mastick bill was approved during the closing hours of the final session. The bridge would be built without cost to the state, as a toll bridge. The last sentence of the article confirmed that the Poughkeepsie project bill died in committee.[3]

A year later, in the May 1923 issue of the *Scientific American*, an article written by William A. Welch appeared as the construction of the bridge was just beginning. Welch offered an exhilarating and detailed preview of what was planned:

> The new bridge will consist of a main span of 1,623 feet in length between the center of the main towers, and two shore spans, each 210 feet in length. The floor of the bridge will be carried on two 17-inch wire cables attached to solid anchorages at each end and carried upon two towers each 350 feet in height, measured from the top of the concrete piers. The underside of the bridge will have a clear height of 153 feet above the river. The traffic will be accommodated on a single floor which will consist of a central roadway 30 feet in clear width between the curbs and two sidewalks, each four feet in width. The floor will be hung from the cables by 2-1/2-inch wire ropes. At their lower ends the suspenders will be attached to transverse floor beam trusses, the distance between the trusses being spanned by longitudinal plate steel stringers. Upon this system will be laid a roadway consisting of a concrete base with an asphalt block surface. To stiffen the floor under the moving traffic two stiffening trusses 30 feet in depth are provided, one on each side of the floor. These trusses will be spaced 55 feet apart, center to center. The triangulation of the trusses will be the same as that of the Manhattan Bridge across the East River, with non-intersecting members and verticals at each angle. Each tower will consist of two built-up posts tied together with stiffening trusses. Each post is 30 feet in width at the base, measured parallel with the axis of the bridge, and tapers to a width of 11 feet at the top. Transversely, each post has a width, throughout its full height, of 6-1/2 feet. Although the posts have a rather light appearance for a bridge of this magnitude, each is a very stiff member, consisting of two closed double-box outer sections, tied together on their inner faces with a heavy plate diaphragm and with two lines of latticing.[4]

Welch was chief engineer at the PIPC and served as general manager for the commission as well. Once the project was in full swing, he served as a consulting engineer, overseeing and approving changes in the design and construction on behalf of the Palisades Interstate Park Commission and the state of New York.

Another article appeared on March 23, 1922, in the *Bernardsville (NJ) News* that reported on the impact that the construction of the bridge would have on the local economy as well as the far-reaching overall effects. The article projected that "ten thousand carloads of material and the labor of five hundred men for two years" would be required if the bridge moved forward. Tench himself authored the piece, confirming the immense impact that "such undertakings as the erection of a bridge of this sort (usually) have on unemployment."[5]

Additional details about the large quantities of steel, lumber, concrete, and other materials stirred interest. For instance, it was estimated that at least three thousand tons of concrete would be needed, and for each ton of cement, quarrymen would have to mine four tons of rock for crushed stone and two tons of sand. Additionally, a large amount of dynamite would be required to blast out the anchorages and the roadway along Anthony's Nose. There was also the human element: not only the men needed as laborers for the project but also those individuals involved across the nation in manufacturing the building materials and transporting these materials to the site. Additionally, the cost of hiring five hundred men to work on site for two years along with the cost of off-site work such as engineering and design was estimated to be around $5 million.[6]

On March 18, 1922, the Smith-Mastick bill—so named for its cosponsors, Seabury C. Mastick and C. Ernest Smith, of the New York Senate—officially made its way through the New York Legislature and was signed into law by the governor of New York. A charter for the Bear Mountain Hudson River Bridge Company (BMHRBC) was approved and incorporated, and the construction of a toll bridge across the Hudson at the location noted was granted.[7]

Besides the state approvals, Tench and his sponsors had to seek federal approval by sending an application to the Department of War,[8] which was required for any proposed bridge that impacted waterway access anywhere within the country. The federal government still mandates the same review process even today to be sure that waterways remain unobstructed. This regulation was introduced in 1906, as industrial development was accelerating throughout the nation. Congress enacted legislation ensuring that "to construct and maintain a bridge across or over any of the navigable waters of the United States, such bridge shall not be built or commenced until the plans and specifications for its construction . . . have been submitted to the Secretary of the department in which the Coast Guard is operating for the Secretary's approval, nor until the Secretary shall have approved such plans and specifications."[9] This language was adopted to guarantee the free navigation of waterways and ensure the present navigable depths would not be impaired, nor the channel or channels through the structure be obstructed, thus affecting commerce and national security. The legislation was codified under the "Act to Regulate the Construction of Bridges over Navigable Waters" (33 USC, 491). On May 25, 1922, Congress issued an act approving the application for the Bear Mountain Hudson River Bridge. The federal legislation required the bridge to be started within three years and completed within seven years of that date.

Now that the approvals were issued, the Bear Mountain Hudson River Bridge Company not only had the monumental task of raising sufficient funds to finance the construction, but also faced the organizational challenges of setting up and carrying out a business venture of considerable scale. Correspondence dated November 1922, between Roland, who had agreed to serve as the president of the Bear Mountain Hudson River Bridge Company, and George Adams Ellis, legal counsel for W. A. Harriman & Co., makes the enormity of the task clear; the men discussed the structure of the new company as laid out in the charter. Ellis, a partner in the firm of Clark, Carr, and Ellis, set down his analysis and suggested some changes that he felt were necessary. They were based on a study of the charter document undertaken by Ellis and several members of his firm. In his letter, he noted that he had reviewed the certificate of incorporation and other corporate records for the BMHRBC and was "entirely satisfied with the form of these records so far as the essential points are concerned."[10] However, he suggested some minor modifications related to financing, the form of the construction contract, and securing land and rights-of-way.

The original financing structure involved securing approximately $3 million from first-mortgage bonds, which would yield $2.7 million in cash; $1 million from debentures or second-mortgage bonds, which would yield approximately $900,000 in cash; and $312,500 in cash from capital stock, secured by issuing 12,500 shares of stock, at $25 per share. Ellis noted that based on the estimates submitted, this total of $3,912,500 left the company approximately $500,000 short of what he projected would be the required amount. His suggestion, which he felt was the only practical solution, was to increase the first-mortgage bonds by $500,000, that is, to an amount equal to the actual construction cost. While his advice was firmly presented, he, of course, left the final decision in the hands of Roland and his board of directors. According to the company charter, issuing bonds was authorized provided that a sinking fund was established and the plans for the fund were approved by the Public Service Commission of the state of New York. That commission would designate the banks where the money would be deposited according to the state's rules and regulations.[11]

The plans and specs for the construction, prepared for Tench by Howard Carter Baird of Hodge & Baird, would need to be reviewed and approved by the state engineer and surveyor as well as by the Palisades Interstate Park Commission. According to the state legislation, construction was to begin within one year of the date that all consents were officially granted by the federal authorities (August 21, 1922) and completed within three years of that date. The state authorization issued a timeline that slightly differed from the one outlined above in the federal legislation. The Smith-Mastick bill also authorized the acquisition of lands and other property by condemnation. The BMHRBC would be required to file statements listing the cost of any acquired land with the state engineer and surveyor as the project progressed. Various restrictions on building, financing, and operating imposed by the federal and state governments and the corporation's bylaws, some of which overlapped, added additional obstacles to be navigated.

The secretary of war and chief engineers gave their approval with the caveat that the district engineer of the Army Corps of Engineers would be involved in supervising the bridge's construction. Perhaps most important, at least when it came to managing a return for investors, the Smith-Mastick bill gave the state the right to eventually acquire the bridge according to the following schedule:

- 5 Years after completion for $4,500,000 or
- 10 Years after completion for $4,000,000 or
- 15 Years after completion for $3,500,000 or
- 20 Years after completion for $3,000,000 or
- 25 Years after completion for $2,000,000 or
- 30 Years after completion at no cost.

The state also prescribed maximum toll rates but required that rates be dictated and controlled by the Public Service Commission. The company's Certificate of Incorporation, which governed its business practices, was subject to the restrictions of New York's stock corporation law. Accordingly, starting capital had to be fully funded before starting the business, and directors were personally liable for debts incurred.

Per the company charter, eleven shares of the stock were subscribed in the Certificate of Incorporation with no price fixed; in his correspondence with Roland, George Ellis recommended that a price be affixed to these shares and that the subscriptions be paid for. Also, the corporate bylaws dictated that no business be transacted at the annual meeting. Ellis disagreed with this rule too, recommending that this language be amended to allow business to be conducted. Ellis suggested revisions to a few other bylaws and made recommendations regulating the size of a quorum, defining the process required to pass resolutions and elect committees, as well as other governance-related issues. He also recommended several best practices for record keeping and maintaining proper documentation backup (for example, signing off on minutes, procuring written resignation documents when necessary, and so on). The Certificate of Incorporation fixed the principal office of the Bear Mountain Hudson River Bridge Company to an address in White Plains with another location in New York City. Ellis was against listing any location on the certificate. Instead, he indicated that it should allow the office to be anyplace the directors deemed appropriate. The charter listed the following board of directors:

>C. F. deGanahl Wilson Fitch Smith
>Daniel D. Wever George S. Field
>E. Roland Harriman Edward W. Freeman
>Cornelius A. Pugsley Carlton M. Smith
>Edward Mills [one vacancy]
>William T. Smith

The executive committee was made up of Carlton M. Smith, E. R. Harriman, Daniel D. Wever, William T. Smith, and C. F. deGanahl. The officers were Harriman (president), Wilson Fitch Smith (vice president), William T. Smith (secretary and assistant treasurer), and Cornelius A. Pugsley (treasurer).

In communications with Roland, Ellis questioned whom deGanahl, Freeman, Pugsley, Carlton M. Smith, Mills, and Field represented. He also recommended the resignation of Wever, who he knew had been nominated by Terry & Tench, as well as any other directors nominated by that company. Their presence on the board raised a possible conflict of interest since the BMHRBC was considered the owner and would later enter into a contract with Terry & Tench to build the bridge. He also recommended that the remaining vacancy on the board be filled.[12]

Ellis would prove to be an invaluable asset to Roland and the Bear Mountain Hudson River Bridge Company. As principal of Clark, Carr, and Ellis, he had worked with Roland for years and developed a relationship based on mutual understanding and respect. He keenly understood that structuring the corporation correctly was of ultimate importance.

As things moved forward, it became evident that Harriman and Tench disagreed on several points relating to structuring the company and managing the investments necessary to raise funding. The BMHR Bridge Company charter listed E. Roland Harriman as president of the company, but Tench and several members of Terry & Tench Co., Inc., were also listed in the charter as officers of the new company. Roland agreed with Ellis that representation of the contractor on the board of the ownership entity would be a conflict of interest. He presented this matter to Tench. Tench, of course, did not agree. The men also disagreed on certain financing arrangements. The relationship between the two, so enthusiastic just months earlier, had taken a sharp turn. Since they were not able to resolve their differences, Roland resigned and stepped back from the project sometime in April 1922, shortly after the state approvals were issued, and took with him his financial plan and connections. Roland may have assumed that Tench would give in to his point of view considering Roland's financial prowess and his contacts within the business community. This difference of opinion threatened to derail the whole project. Harriman was not easily dissuaded from his opinion. But then, neither was Tench.

16
Terry & Tench

Frederick Tench was a self-made man and one not easily intimidated. He and his partner, Edward Terry, arrived in New York City in 1894 and established their firm: Terry & Tench Co., Inc. They worked diligently, erecting steel for bridges and commercial buildings in the region, and gained great success. By the time Tench proposed the bridge project, to be erected across the Hudson River south of Albany, connecting Anthony's Nose and Bear Mountain Park, the company had a reputation for competence and success. Although others before him had tried more than once to build a bridge in this same location, Frederick Tench, of Terry & Tench Co., Inc., decided that now was the time for it to become a reality—and that he was the man to do it.

The partners had met while working as bridge builders in Winona, Minnesota, years earlier, learning the trade as they perfected their engineering skills.[1] Tench, the more aggressive of the two, had emigrated in 1888 from Grimsby, Ontario, where he lived with his parents, William Eastwood Carruthers Tench and Ellen Murray Tench. His father was a local entrepreneur who ran a sawmill, a flour mill, and a tavern. Frederick was born in 1862, went to the local schools, and graduated from high school, at which time William brought him into the family business as a partner, adopting the business name Tench and Son. Unfortunately, the company met with difficult circumstances, and in 1888 William was forced to close his doors in bankruptcy. Unable to pay his debts, he left many of his creditors unpaid and in bad financial condition. It was a tremendous source of shame and dishonor for the elder Tench, and it bothered him until the day he died in 1895.[2]

After Tench and Son failed, Frederick decided there was little for him in Grimsby, and he left Canada to start a new life in the United States. He worked his way through the Midwest, joining up with crews building bridges across the Mississippi and Missouri Rivers and other lesser tributaries. He traveled across several states, following the projects where they led him, working as a laborer at first, dedicating himself to learning as much about bridge building and steel erection as possible. He worked by day, learning the trade, and studied by night, completing correspondence courses in engineering to expand his knowledge and further his career. During this time Frederick met Edward Terry, another motivated young bridge builder, four years his senior. They became good friends right from the beginning. While Tench was building a bridge over the Mississippi River in Minnesota, Terry was involved with a different bridge over the same river in Wisconsin.[3] They continued to cross paths on bridge projects in Louisiana, Oregon, and Idaho, as they worked tirelessly to increase their knowledge in the hopes of someday striking out on their own. At one point they both worked for the Union Bridge Company and later served as construction superintendents for the Union Pacific Railway for several years.

Edward F. Terry was born in 1857 in Concord, New Hampshire. He was orphaned as a young boy and sent to live with his uncle on a farm in Wisconsin where he learned his work ethic. His fascination with railroads began at a young age, and he spent many hours in the company of the railroad workmen along the tracks listening to their stories. After meeting up with Michael Riney, an old-time bridgeman, who took a liking to him, Terry secured a job as a riveter with the Rochester Bridge Company, where Riney worked. When Riney was injured while working on a viaduct, Terry was given a chance to take his place, and he excelled at the opportunity that had come his way. Thus began his career working on bridges and railroads. Like Tench, he began traveling about the country following available jobs. He was involved in building as many as five crossings over the Mississippi River and several bridges over the Missouri River and other rivers in the Midwest, employed by several different construction companies, including the Youngstown Bridge Company and the Union Bridge Company. Eventually, he landed at the Union Pacific

Railroad, doing construction work on railroads in Idaho and Oregon, as he continued to build his reputation.[4] In the late 1880s, while still in Wisconsin, Terry met Rebecca Allie. They were married in 1889.

At Union Pacific, Tench and Terry worked side by side, confiding in each other about their dream of one day striking out on their own. Considering the state of the railroad, it is not hard to understand why they may have dreamed of greener pastures. At the time, Union Pacific was experiencing major financial difficulties owing to several unfortunate circumstances. In 1880 the railroad had been pushed into a chaotic merger with the Kansas Pacific (and its subsidiary the Denver Pacific). The Kansas Pacific was a poorly built railroad and was badly managed, conditions that resulted in the railroad being put into receivership by 1874. While Union Pacific operated at an annual surplus, both the Kansas Pacific and the Denver Pacific had annual deficits; while Union Pacific did not default on any of its commitments, the Kansas and the Denver did nothing but. To be sure, Union Pacific did not want the merger but agreed to it only as a defensive measure. The Kansas and the Denver were cutting rates to a level only pennies above operating costs. Union Pacific stood to lose much in revenue if it allowed this price gouging to continue and believed the only way to fight the competition was to bring the competition into the company. The absorption of the two lines in 1880 brought along deficits rather than capital, causing the newly formed entity to immediately weaken financially. Between 1884 and 1890, under the leadership of UP president Charles Frances Adams, company finances declined even further when Adams decided to absorb still more branch lines to increase revenue. Although the revenues did increase somewhat, the absorbed lines brought no additional capital to the company, thus diluting the asset/debt ratio still further. In 1892 it became apparent that there would be difficulty paying several short-term notes that were coming due. In addition, the repayment date of UP's government loan—a debt of close to $53 million—was lurching closer, due in 1895, and the capital necessary to satisfy that debt was unavailable. By the time the financial panic of 1893 hit, UP was in a precarious position that would have caused it to struggle in the best of times. Given all these circumstances, it appeared inevitable that in October 1893, the railroad was placed in receivership. Soon after this turn of events, UP's investors approached Jacob Schiff of Kuhn, Loeb, and Co., a well-known financial investment firm, to restructure the entity. Schiff soon found himself in a struggle with Edward H. Harriman for the ownership rights to UP. Schiff was well versed in finance but did not know anything about running a railroad. Unbeknownst to Schiff, Harriman had been quietly buying up available stock and had successfully gained a strong foothold in the company before Schiff realized what was going on. The two men met and decided that they would jointly manage the reorganization. By 1897 the new entity had secured full control of the railroad with the backing of the US government, and Edward H. Harriman was installed as a director and a member of the executive committee.[5] How much of this story was common knowledge at the time we cannot be certain. Whether the unstable environment within the company prompted Tench and Terry to leave and strike out on their own, we do not know; what we do know is that these two young entrepreneurs had long been pondering the chance to secure their place in the world, and in 1894, both of them resolved that now was the right time to make their move. They left for New York City with their life savings in hand, approximately $2,000 combined. The Terry & Tench Co. was officially established, with its first office located at 10 East Forty-Second Street, in New York City.

The first contract awarded to the Terry & Tench Co. after arriving in New York City was for the Mills Hotel on Bleecker Street, erected in 1894. In 1895 they took on the Harlem River drawbridge job for the New York Central Railroad, which allowed them to prove their competence as bridge builders.[6] Through this project, they developed a relationship with the railroad, which became a source for many future contracts. While relatively unknown in the New York trade circles, they forged ahead, successfully winning contract after contract. Within five years their firm was a recognized leader in their specialty of steel erection, and in 1899 the firm was incorporated, becoming Terry & Tench Co., Inc. Their decision to move to New York had proved prescient.

In the late 1890s, young Emma Harriet Roberts arrived in New York City to study drama and elocution. Emma was the daughter of prominent Philadelphia attorney John Roberts and his wife, Sarah J. McKernan Roberts. Shortly after arriving in New York, Emma met Frederick Tench while both were traveling on an elevated subway train. It wasn't long before Frederick decided that Emma was the woman he wanted to marry, and he proposed sometime in 1897 or 1898 during a visit to Grant's Tomb, a popular landmark in Upper Manhattan.[7] The couple was married in 1899 at Calvary Episcopal Church in Germantown, outside Philadelphia, by the Reverend J. DeWolf Perry, the rector of the church. Following the ceremony, a private wedding breakfast was held for the immediate families at the home of the bride's parents. It was reported that Mr. and Mrs. Tench left shortly afterward for an extended honeymoon and planned to reside at the Loyola in New York City.[8]

One year later the New York City Census of 1900 shows them living at Central Park West with one house servant. Shortly thereafter, in 1901, the couple welcomed their first child, John Roberts Tench. In 1905 a second child, Sara(h) Ellen Tench, nicknamed "Nellie" by her family, was born. Frederick and Emma decided that the suburbs would be a better place for a growing family. Sometime around 1905, the family moved to White Plains, where they found a small house that had room for the children to play yet was close enough to the city for Frederick to commute. Their third child, Frederica, was born in 1908. With a still growing family, the Tenches relocated to a larger house in White Plains. This time they found a large frame Victorian home located at 20 Greenridge Avenue.

The house was spacious and comfortable and included a wraparound porch and a wine cellar that housed Frederick's prized wine collection. The family employed two house servants at this time. The grounds included tennis courts, a lovely carriage house, and a garden where Frederick enjoyed puttering. By all accounts, Frederick was a warm and loving husband and father who enjoyed spending time with Emma and the children and Topsy, the family's cocker spaniel, who claimed Westminster Kennel Club bloodlines. A devout Episcopalian, Frederick was instrumental in starting a new church in White Plains, St. Bartholomew's Episcopal Church, where he served as vestryman from the founding of the church until he died.[9]

As Frederick Tench's children grew, so did his business. And with his company doing so well financially, he set out to right the wrongs that had troubled his father and himself for many years. He began with a search for any of his father's creditors who were still in existence. He placed ads in Canadian newspapers and invited creditors of Tench and Son to submit a statement of accounts as they existed at the time of the bankruptcy. He responded to each of them, paying them in full what they felt they were owed, including interest.[10]

By the early 1920s, the Terry & Tench Co., Inc., had completed many high-profile projects, which included the steel framing for such buildings as the Grand Central Terminal, the Biltmore Hotel, the Sherry Building at Fifth Avenue and Forty-Fifth Street, as well as a sixteen-story office building at Pearl Street and Broadway. They bid on and secured contracts for the Williamsburg Bridge spanning the East River in 1903, as well as the towers and span of the Manhattan Bridge in 1908. In 1914, Terry & Tench Co., Inc., had expanded its influence into New Jersey, across the Hudson River, when Tench was elected president of the Meadows Taxpayers Association, a group of companies involved in developing land along Passaic Avenue and the Old Plank Road in Kearney, New Jersey.[11] Terry & Tench Co., Inc., also helped build the New York subway system, securing contracts for multiple subway tunnels and elevated rails; they laid all the rails for the lines above 104th Street and constructed railway viaducts and bridges for both the New York Central Railroad and the Brooklyn Rapid Transit Company.[12]

In New York and across the nation the influence of labor unions and the growth of the labor movement were increasing. It is difficult to determine how much of the work under construction at the time was considered union and how much was done as nonunion. It would be plausible to assume that Terry & Tench might have supported both union and nonunion projects. There is evidence that they did work with the unions, since we know that they were awarded the contract for the towers of the Manhattan Bridge over the East River in 1908, and information gathered from the archives of the Iron Workers Local 361 for New York City indicates that the Manhattan Bridge towers contract was a union contract.[13]

The ironworkers (also known as "bridgemen") of the time were blue-collar workmen, rough and strong, displaying a "no-nonsense" attitude and a sense of loyalty to their brother bridgemen that made them resistant to trusting anyone outside of their profession. Transcripts of interviews with ironworkers at the time conducted by union officials provide some insight into the men and their mind-set. The comments noted below are the responses of an ironworker to questions posed regarding conflicts on the job:

> I ain't got nothin to tell ya. . . . We don like to talk around here. We got our own worries. Ask Mike over there. He'll tell ya lotsa stories. Hey Mike! Here's a feller wants ya to tell him stories. Ya see he ain interested.
>
> Naa the men don feel like talkin. . . . We don't have those kind of guys around here. . . . They never talk. . . . They aint interested in it. They'll laugh at ya. . . . There ain't nothing in it for us.[14]

Of the many trades counted within the construction industry, the ironworker was the most dangerous job—men walked on narrow beams at extreme heights carrying materials, tools, and so forth, and at the time there were no safety regulations in place; nor were there government agencies to enforce compliance with regulations, as we have today. Statistics note that in 1902 the annual average of deaths in the trade was 12.5 percent. The ironworker's motto at the time, "We do not die; we are killed," was a defiant response to the constant threat of injury and death that they faced every day. Whether union or nonunion, the men shared the same values of hard work, camaraderie, and strong determination to get the job done. Those workers who were union members would have confirmed their affiliation by registering with Local 361 out of Brooklyn and paying their monthly dues. Those ironworkers considered nonunion would have had few differences other than not declaring membership.

The Second Industrial Revolution occurred in the United States soon after the end of the Civil War. The country experienced a new, higher level of sophisticated technical capability in manufacturing as well as new inroads in transportation, with the latter having a positive effect on fighting the labor shortages that were prevalent at the time. The introduction of railroads and steamboats allowed workers to travel from state to state more easily, following available jobs. This easy movement across the country changed the distribution of the population as cities experienced an increase and rural areas a decrease. In addition, the United States welcomed twenty-five million immigrants between 1865 and 1910, which impacted the population count as well as the availability of labor.

The average unskilled laborer in New York City was at the mercy of businessmen who owned and managed the companies that employed them. Industrial titans such as J. Pierpont Morgan, John D. Rockefeller, and others set standards that favored management with little recognition given to the worker. The first organization of ironworkers recorded in New York City was in April 1886 when some German immigrants opened an office in Brooklyn. Their concerns were about workers' rights and safety standards. As time went on, there were multiple attempts at creating a system of centralized leadership through the development of various national labor organizations, but there were many difficulties. In 1896 the International Association of Bridge and Structural Ironworkers was formed, and the Bridgemen Local 2 covering New York City and Brooklyn at the time joined it. The *Bridgemen's Magazine* (now known as the *Ironworkers Magazine*) was then the communications link for the International Association with the member unions it managed. Despite many attempts to organize, problems continued, and in 1904, several issues occurred between Local 2 and the International, prompting a reorganization of Local 2 into what became known as Local 35. Years later, organizational problems again developed and resulted in Local 35 losing its charter. In August 1920, the local was reorganized once more into Local 361, which services the New York boroughs today. As the new Local 361 attempted to find its footing, stories of disorganization and radicalism continued to circulate, and the local gained a reputation for extremism and intolerance.

In 1902 the standard wage for an ironworker was $3.50 per day, based on nine hours of work (which translates to $0.3888 per hour). By 1921 wages were recorded to be $1.125 per hour, but union workers were required to pay $2.00 per month in union dues. Many employers and contractors would not negotiate with unions, and much of the work was done nonunion. A strike ensued by structural unions against the New York City Iron League and the National Erectors Association (two organizations formed by contractors and owners). The strike lasted from May 1, 1924, through the next fourteen years.[15] The Bear Mountain Bridge was built from 1923 through 1924. It does seem that the Bear Mountain Bridge was a union project, probably just missing the impact of the strike. We uncovered a newspaper article in a local paper recounting that on June 14, 1924, at the Bear Mountain Bridge, several bridge workers went on strike, and ownership ordered six state troopers of the state constabulary to be stationed at the project, as a precaution against any possible violence developing that could result in damage to the new bridge.[16] The bridge workers claimed there was a wage dispute and that the contractor owed them back wages. They said they had been laid off after completing the more complicated work, and the contractor was bringing in nonunion labor and paying them less money.[17] How the situation was ultimately resolved we do not know, but it confirms that the bridge workers were union men. There is information that supports the fact that Local 361 kept working during this time, but records show construction in general was slow.

Sometime in the first decade of the twentieth century, Terry and Tench met up with Olin S. Proctor, the inventor of a tunneling machine. Proctor seems to have come up with the original design and initially applied for the patent. It is unclear at what point Terry and Tench became involved in the enterprise; perhaps they collaborated with Olin on the machine's design and details, or maybe they were consulted for their financial backing and business acumen to organize and market the idea. Regardless, Proctor brought Terry and Tench in on the project at some point, and all three men assumed ownership of the product, which was purported to be able to "eat tunnel rock in bites per minute." The US Patent Office references application No. US41637408A as having been applied for by the Terry, Tench, and Proctor Tunneling Machine Company on February 17, 1908; Patent No. US900951A was granted to that entity on October 13, 1908, and published on the same day.[18] The machine churned through rock at a rapid pace, enabling the operator to dig tunnels more quickly, easily, and cheaply than any of the competition. In 1910 full-page advertisements appeared in various local newspapers inviting the public to invest in the venture and join the men in an incredible business undertaking. "Buy stock and become a partner," the invitation boomed, "in the Wonderful Machine!"[19] The outcome of this venture is not clear. What is clear is the versatility of the partners and their endless energy and ingenuity.

Terry and Tench's reputations as competent, professional engineers continued to grow, as reports of several very complicated projects the company had successfully taken on appeared in New York–area newspapers and traveled by word of mouth within the industry. One such project was moving an entire section of a steel tunnel frame between 135th and 136th Streets, an amazing feat of engineering that took several days of tedious, precise planning. The block had been completed as it was designed to accommodate a two-track railroad. Terry & Tench Co., Inc., got involved when the Rapid Transit Commission (the precursor to the MTA, the Mass Transit Administration) changed the plan, ordering that the frame be adjusted to accommodate *three* tracks. The company moved the entire fully constructed roof assembly and one side of the framing, sliding thousands of tons of steelwork far enough to allow for the additional rail. The project was a success with no damaging effects on the existing structure.[20] That was not the only seemingly impossible task they accepted: they also managed to move a 265-foot double-decker bridge, spanning the Harlem Ship Canal, relocating it from 215th Street to 231st Street. The sixteen-hundred-ton structure required only thirty-five minutes to move, but the time and effort that were necessary to prepare for the move and to set the span into its new housing caused traffic to be interrupted for two full days.[21] Again, the project was successful with no damage done to the structure.

Their talent seemed to have no boundaries. Nor was their work limited to contracts for steel framing in bridges, tunnels, and buildings in New York. There were other areas in which they dabbled. When the United States entered World War I, the partners, like many businesses at the time, wanted to serve their country in the best way they could. The Terry Shipbuilding Corporation was established, and a contract was drawn up with the government to provide twenty ships, each three thousand tons, built of steel and wood, and powered by steam. For its shipyard, the company purchased land from the Port Wentworth Terminal, located on the Savannah River in Savannah, Georgia. Frederick Tench reported in a newspaper interview that the site was chosen because of its direct connection with the Savannah and Northwestern Railroad, the abundance of local labor, and the ready availability of shipbuilding materials in the area. Since the expected schedule for the ships to be delivered was tight, another important factor was the mild climate that meant the builders could work year-round. The ships were completed and delivered well within the required time frame.[22]

Terry & Tench Co., Inc., was also awarded other US government contracts. They not only built several steel towers for the US Navy in Annapolis but also remediated the structural framing for the Lincoln Memorial in Washington, DC, before its opening.[23] Ground was broken, and another contractor did the initial work on the memorial starting in 1914. In June 1921, Terry & Tench Co., Inc., signed an agreement with the federal government to do some remedial work on the structure. It was determined that the memorial needed structural reinforcement. The company installed additional underpinning for the approaches as well as for the terrace walls. This work was accomplished by carrying the foundations to the bedrock. The contract was for $220,359.28. The following January, another contract was issued to them to repair concrete damaged by settlement, this one for $63,408.92.[24] The Lincoln Memorial was finally ready, and it opened in 1922, eight years after its groundbreaking.

In 1922, when Tench approached the Harrimans with the Bear Mountain Bridge project, Terry & Tench Co., Inc., was at the peak of its success.

While assisting Tench in coordinating the Bear Mountain Bridge project, Roland spoke to several business associates with whom he and his family had relationships. These successful businessmen were always looking for a good investment; at the same time, they were supportive of the community. Conversations between Roland and these men evolved into a full-scale financial arrangement that would eventually provide the funding necessary to build the bridge. Building the bridge was an experience that Roland did not initially seek out, but he welcomed it wholeheartedly, as he did all his life adventures. Indeed, such an exuberant embrace of risk taking was a central theme in his memoir, *I Reminisce*; after describing each new experience he embarked upon, he noted just how educational and enjoyable it was. His reminiscence on the Bear Mountain Bridge project was no different: "I learned an awful lot, and I think I did contribute to the good of the country, and I had a lot of fun, as usual."[25]

Being Henry's sons, Roland and Averell were involved with railroads from an early age. As early as 1920, Roland was a director of the Delaware and Hudson Railroad. After Henry died in 1909, his associate and good friend Judge Robert Lovett took over the leadership of the Union Pacific until he died in 1932. Averell subsequently became chairman of the board at UP, and Roland, along with Judge Lovett's son, Robert, became directors. Averell eventually found himself immersed in a career of political diplomacy and foreign relations during the war years and afterward. This turn of events led to him ending his connections with the railroad.[26] As he exited his responsibilities at Union Pacific, Roland succeeded him and was active as chairman until 1968, remaining on the board for the next ten years until his death.[27] The brothers' different personalities complemented each other, with Roland playing the role of the tortoise to Averell's hare. Roland was deliberate, detail oriented, and conservative in his social preferences; Averell, on the other hand, was aggressive and stimulated by pressure, having inherited his father's passion for business. Averell loved to travel and was comfortable in the limelight. Roland, more unassuming and patient, much preferred his quiet life with Gladys.[28]

That said, Roland had diverse interests over the years, and his business ventures were many and varied. In 1931, W. A. Harriman and Company merged with Brown Bros. to form a new investment company called Brown Brothers Harriman and Co. Roland maintained his position in the firm from the time of its creation. However, he often pursued other interesting opportunities simultaneously. In 1923 he invested in the magazines *Time* and *Newsweek*, and for a while, he was also involved in running a local newspaper that he had acquired in Middletown, New York.[29] Throughout his life his many investments and interests introduced him to a variety of pursuits besides his main vocation of finance, everything from railroads to shipbuilding, publishing to bridge building, philanthropy, the Red Cross and the Boys Clubs, to horse racing and structuring the United States Trotting Association. And through it all, he maintained that he continued to learn and to have fun while doing so.

17
A Difference of Opinion

The bridge project intrigued Roland, but he was concerned that all appropriate standards would be met in structuring the corporate entity, both legally and professionally. His discussions with Frederick Tench regarding the recommendations set forth by George Ellis, his legal counsel, seemed to go nowhere. Tench was adamant that he and his associates should remain as officers of the ownership entity, regardless of the concerns that doing so would create a conflict of interest. Roland would not participate in the project if they could not reach an agreement concerning the matter—therefore, Roland walked away.

Since Tench had already been introduced to several of Roland Harriman's associates (by Harriman himself), he accepted Roland's resignation and decided to petition George W. Perkins Jr., a supporter of the bridge and a commissioner of the Palisades Interstate Park Commission, to step in as president of the new Bear Mountain Hudson River Bridge Company. Perkins, however, was out of the country on business for G. Amsinck and Company, Inc., an import-export company, when Tench sent his offer.

One of Perkins's colleagues at Amsinck, identified only as "Ed," wrote to Perkins on May 4, 1922, to discuss Tench's offer. It is assumed that Perkins and Ed had a previous communication about the matter and Perkins had asked Ed to do some checking for him. Ed explains that he had talked the matter over with Perkins's associates per Perkins's directive. Both Major William A. Welch and John Hall were mentioned as two colleagues whose opinions Perkins valued. Ed reports that Welch "felt quite strongly that it would be a very desirable thing both for you and for the Park if you would serve in that capacity." Welch was the general manager of the PIPC and, of course, was considering the impact of the arrangement on the agency.

Ed wanted to give Perkins all the relevant information so that he might review the matter during his travel home. In his letter, he also updated Perkins on several events (and his impressions of them) that had occurred since the twenty-first of April, when the BMHR Bridge Company met and voted to elect Perkins to the role of president (absent his consent):

> Roland Harriman has been in touch with me and seems to be somewhat dissatisfied with the situation; in fact, so much so that he tells me that he has tendered his resignation as a director of the company. He bases this action on what he terms "a dangerous relationship" between the contractor and the bridge company and also, on what he considers unpractical financing arrangements. . . . I am inclined to feel that his criticisms are unduly severe and his fears without real foundation, but . . . I am asking Mr. Tench to send me copies of all proceedings to date and I am going to go over these very carefully, together with . . . John Hall and Major Welch with a view to deciding whether or not the situation contains the unsatisfactory elements which Roland has mentioned. I am not at all sure that his opinion is not in some measure influenced by the fact that negotiations for financing between Tench and W. A. Harriman & Co., Inc. were broken off rather abruptly when Tench refused to seriously consider some rather radical changes in his plans which the bankers suggested.[1]

Unfortunately, no additional communications on the subject between Perkins and Tench or between Perkins and Harriman seem to be available. But a letter from the same Amsinck associate, Ed, dated May 16, 1922, confirmed to Perkins that Ed had received the papers he had requested from Tench and that he had conferred with Welch as he noted in his previous letter. Ed also confirmed that the relationship between the contractor and the bridge company was very close and offered justifications for this arrangement. For their part, Hall and Welch both recommended that the bridge company employ different counsel from Terry & Tench.[2]

Another concern raised by Roland as well as by a board member, W. T. Smith, was that it was dangerous to award a contract for the bridge before complete and detailed plans and specifications had been prepared and adopted. Concerns about cost overruns occurring once the project was under way were also discussed. Eventually, it was agreed that work could not be started until the state engineer, the Palisades Interstate Park Commission, and the consulting engineer of the BMHR Bridge Company reviewed and approved the plans and specifications. Additionally, Harriman, Smith, Hall, and Welch all agreed that the best protection against excessive costs would be a performance bond. Terry & Tench agreed to furnish such a bond for $1 million; they felt it was an appropriate amount to protect the company and guarantee the continuity of the project.

In the same letter of May 16, Ed references a conversation he had had with Roland during which Roland stated that he was not "thru [sic] with The Bridge Company by any means and hoped to get back into it in some capacity or other." Ed also disclosed: "Mr. Tench seems rather peeved at Roland's attitude. He says Roland has known all along that Terry & Tench were in this thing primarily to get a contract to build a bridge; that Roland has been thoroly [sic] informed of each step that has been taken and has had ample opportunity to comment or criticize. Mr. Tench does not seem to be contemplating any move to revive Roland's association with the project."[3]

This correspondence mentioned several attachments of additional letters that may have provided further details on the financial arrangements made up to that date, along with the names of several investors who were friends and associates of Terry & Tench; Ed referenced letters from both Roland and W. T. Smith as well.[4] Unfortunately, none of these attachments could be located, so we were not able to review them. Consequently, we are left to wonder what were the financial issues that existed at this time and why alternative arrangements were proposed to resolve them.

According to the documents we were able to uncover, the arrangement was never finalized. Whether Perkins declined the appointment for his own reasons or whether Roland Harriman had something to do with his final decision, we do not know for sure. However, according to sparse correspondence dated within this time, it seems that Tench and Harriman ultimately resolved their differences. In a letter from Roland to Tench, dated June 12, 1922, Roland again declined to involve his family members and associates in the project "as long as we strongly disapprove of the present methods." A few months later, correspondence between the men indicates that Roland agreed to resume his role as president and take on the task of financing—but only if it was done according to his terms.[5] These demands included requiring that Terry & Tench Co., Inc., remove itself from the BMHR Bridge Company (as originally recommended by George Ellis, W. A. Harriman and Co.'s legal counsel), with the understanding that they would act as a contractor and build the bridge for a fair price only if Roland could secure the financing. Regardless of how the agreement came about, the outcome was that Roland returned to the position of president and Tench accepted his terms.

With the conflict resolved, Roland set to work to sell the stocks and bonds to those individuals in his circle who he thought would be interested. As originally planned, Roland created a financial program structured around selling shares of stock in the bridge to financiers within his circle of associates and other community-minded businessmen. Members of the Harriman family, as well as George W. Perkins Jr. and members of the Perkins family, were all very supportive of the project and invested heavily in the program. As the framework for the company and the necessary components were put into place, the required plans and specifications were reviewed and finalized. Terry & Tench Co., Inc., submitted the necessary estimates, and the parties entered discussions regarding the issuance of a contract to be drawn up between the BMHR Bridge Company and Terry & Tench Co., Inc. As part of the contract, the BMHR Bridge Company would require Terry & Tench Co., Inc., to provide a performance bond to guard against unforeseen financial issues. (A performance bond is a type of surety bond given by an insurance company to ensure the proper completion [performance] of a project by a contractor. It provides the money that would allow the contractor to finish the project should he experience financial difficulties during the project.) The contractor agreed and contacted a surety company to secure the bond. However, Globe Insurance, the surety company, required a cosigner to act as a guarantor for the loan. George W. Perkins Jr. agreed to act as guarantor to the bonding company for Terry & Tench Co., Inc.'s performance bond, a decision he may have ultimately regretted.

Roland's plan to encourage investors to join the project involved writing to General Cornelius Vanderbilt III, explaining the project and inviting him to invest. With the invitation, he included several estimates of cost for the proposed construction along with a projection of operating costs for the bridge, the latter of which had been prepared based on expected traffic figures calculated independently of the income requirements. The income, of course, would be directly dependent on the amount of traffic crossing the bridge, since it would be generated by the tolls collected. The estimates presumed that the revenue would take care of the income bond and sinking-fund charges on both classes of bonds, leaving a small profit that could then be applied to the common stock. In his letter, Roland was candid with the general. He referred to the first-mortgage bonds as investment-class offerings, but admitted that, in his opinion, the debenture bonds "must be underwritten by men who realize the need for a bridge, the great service it would render not only the whole State but the Palisades Park in particular." Roland went on to list several men known to the general who had, by this time, joined in underwriting a total of $1 million of debentures. The list included the following prominent businessmen from the area:

Gates W. McGarrah: banker; served as president and chairman of the board of the Bank for International Settlements (in Basel, Switzerland) and as a director of the Federal Reserve Bank of New York; also was director of more than a dozen industries and banks at one time; philanthropist and community servant.

Frank Munsey: American newspaper and magazine publisher; banker, political financier, and author.

Mortimer L. Schiff: American banker; son of Jacob H. Schiff of Kuhn, Loeb, and Co. investment bankers (who partnered with E. H. Harriman in Union Pacific Railroad); was himself a partner in Kuhn, Loeb, and Co.; founded the Equitable Trust Co.; served as president of the Boy Scouts of America; philanthropist.

Charles H. Sabin: banker; Democratic supporter of Franklin D. Roosevelt and Alfred H. Smith; was against Prohibition and worked for its repeal.

F. D. Underwood: president of Erie Railroad and director of Wells Fargo and Co.

L. F. Loree: civil engineer, lawyer, railroad executive; has been credited with coining the phrase "This is a helluva way to run a railroad."

B. B. Odell: American businessman and politician; was the thirty-fourth governor of the state of New York (1901–4).

C. A. Pugsley: Democratic congressman from the Sixteenth District; resident of Peekskill; banker and president of the New York State Bankers Association; created the Pugsley Medal that was awarded to champions of parks and conservation.[6]

Once the newly formed company was structured, Terry & Tench submitted an updated estimate, dated November 6, 1922, that proposed a cost of $3,653,000 to be paid through $20,000 in common stock, $225,000 in debenture bonds, and the balance in progress payments.[7] January 1, 1923, was proposed as the bridge's start date and January 1, 1925, as the completion date. This document was used as the basis for the construction contract that was issued between the Bear Mountain Hudson River Bridge Company and Terry & Tench Co., Inc. The contract was drawn up, but the parties involved awaited the okay from Ellis indicating that funding was finalized before they would move to execute it.

In a letter from George Ellis to Roland, dated July 3, 1923, the attorney provided a draft of a letter he suggested Roland send to Tench detailing the procedure that would be used in processing monthly payments against the contract. Standard practice dictated that at the end of each month, a full accounting would be submitted by Terry & Tench to BMHR Bridge Company, listing the value of each area of work, broken down by trade, and the percentage completed as of the date of that invoice; the values were to be calculated using unit pricing provided by the contractor at the start of the project. Each month the total value of work completed to date would be calculated, and the payments delivered as of that time would be listed and deducted, leaving the amount still not completed and therefore still unpaid to the contractor as of that month. Throughout the project, the contractor would receive only 85 percent of the completed value. A retainage of 15 percent would be held back by ownership until project completion, at which time it would be released to the contractor, but only after all items of work were inspected and accepted.[8]

As the company worked to put the necessary paperwork in place, Roland pursued investors, working diligently as a salesman (a task he typically did not perform) for the project during January, February, and early March 1923. In his memoir, *I Reminisce*, he offers some context for why he agreed to such a role and some of the twists and turns it took:

> A man by the name of Tench, a partner in a firm of engineers named Terry and Tench, came to see me with a plan to build a vehicular bridge across the Hudson River. This was in the days when the only crossing of the Hudson south of Albany was by ferryboat. . . . Since I lived on the west side of the Hudson the idea appealed to me immensely. . . . So the financing of the Bear Mountain Bridge, such a new kind of enterprise, was another matter. I thought that first mortgage bonds could probably be sold to the public, as a speculative investment, but the equity money was something else again. The possible purchasers of this higher-risk item were limited to public-spirited citizens who understood the situation and were willing to make a big bet. . . . Well, this was the first time I really had a salesman's job, and I called on an awful lot of people. I even made a trip to Daytona Beach, Florida, where John D. Rockefeller was wintering. I never got to see him, but I did see Raymond Fosdick, his man of affairs, and he subscribed to the venture. In fact, I did succeed in getting the financing done, and then to my horror found that the price of steel had increased beyond the amount of money we had raised.[9]

In a letter to E. J. Berwind (another potential investor) sent shortly after he wrote to General Vanderbilt, Roland outlined the same proposed investment particulars he offered the general: "$3,000,000 in 7% First Mortgage bonds, $1,500,000 in 8% Income debenture bonds, 12,500 shares no-par common stock. Each Income debenture bond will be accompanied by 5 shares of common stock @$10 per share for a total price of $1,000/unit." He went on to project annual expenses (which included interest and sinking-fund charges on both types of bonds, maintenance and operating costs, taxes, and the like) to be approximately $634,000, with the annual toll income based on 652,000 cars (at the time considered to be a conservative projection) at a fee of $1 per car, coming in at $652,000; this projection left a profit of $18,000, which could be applied to the common stock. Roland acknowledged to Berwind, as he had to Vanderbilt, that the income debenture-bond investments were speculative and might be risky, but he appealed, again, to the prospective investor's support of the state and the valley, pointing out that "the bridge will render a large service not only to the whole state but to the Palisades Park in particular."[10]

By February 1923 the list of investors had grown even more and included several men known to both Vanderbilt and Berwind. Besides those persons mentioned in the letter sent to the general, Roland added several new investors, including:

> *Thomas Cochran*: banker; football coach for the University of Minnesota; vice president of Astor Trust Co.; partner of J. P. Morgan & Co.
>
> *William Pierson Hamilton*: great-grandson of Alexander Hamilton; married to J. P. Morgan's daughter; partner in J. P. Morgan and Co.
>
> *George F. Baker Sr.*: American financier and philanthropist; at the time of his death was the third-richest man in the United States; founded the First National Bank of the City of New York.
>
> *Paul M. Warburg*: American investment banker; second vice chairman of the Federal Reserve under Woodrow Wilson.

Roland's unrelenting canvassing of prospective investors yielded success, and by the end of February 1923, just $300,000 remained of the $1.5 million in income bonds on offer. However, because the banks would not commit themselves to the first mortgage until the syndicate on the income bonds was completed, it was important to move forward on it as quickly as possible.[11] Time was slipping away. The BMHR Bridge Company had to begin construction soon to meet the desired completion date. The positive response to Roland's enthusiastic selling continued, fortunately, and by April 20, 1923, subscriptions had been received for the entire issue of income bonds and shares of common stock.[12]

In a letter to Roland, dated March 22, 1923, George Ellis noted that he had in his possession signed right-of-way agreements from the Hazard family, the Hazard Estate, and the Van Cortlandt family, owners of land parcels that were needed to move forward with the bridge construction. Additionally, he had approvals from the Armory Commission and the Palisades Interstate Park Commission, both of which agreed to protect the company as regards rights-of-way. Ellis also informed Roland that he had met with Commissioner Semple of the Public Service Commission and secured Semple's assurance that there would be no obstacle to the company eventually obtaining formal approval for the sinking-fund provision of the bonds, even though it had not yet been granted. This move was meant to reassure Roland (and the other investors) that it would be okay to begin construction: "While it would have been preferable to have had formal approval prior to the execution of the construction contract and definite subscriptions to the bonds, nevertheless, I believe that both the Company and the subscribers are justified in assuming this slight risk incident to obtaining formal approval. . . . [T]he Bridge Company is justified, in my opinion, in executing the form of construction contract upon receipt of an approved form of surety company bonds."[13]

After this approval from legal counsel was received, the construction contract between the BMHRBC and Terry & Tench Co., Inc., was executed on March 24, 1923. Construction commenced in April.

On May 10 of that year, an article appeared in the *Engineering News-Record* written by Wilson Fitch Smith, the project's chief engineer and building supervisor, offering his building-industry colleagues an introduction to his exciting new project:

> The nearest practical location to New York City is at the entrance to the Highlands, about three miles above Peekskill, and forty-five miles from Times Square, where the rocky shores of the river approach within 1,600 ft. of each other. This site was selected thirty-five years ago, by the late Edward W. Serrell for a suspension railroad bridge, and some work was done at that time by the Hudson Suspension Bridge and New England Railway Co. The site is ideal for a suspension bridge. On the east side of the river, Anthony's Nose rises abruptly to an elevation of 900 ft., a solid base of granite gneiss, through which the New York Central R.R. tracks pass in a tunnel. On the west side the base of Bear Mountain, upon which can still be traced the remains of Fort Clinton, is of similar rock formation. Between these two rocky shores, the river has a channel depth of 130 ft. Suitable rock foundations for the towers exist on both sides of the river at the shoreline. Adequate anchorages on both sides can be had by tunneling into the solid rock online with the cable

backstays. The elevation of the surface on both sides is advantageous for the support of the approach spans.

These natural conditions, together with the fact that on the west side of the river lies Bear Mountain Park, a state reservation of over 33,000 acres, an increasingly popular playground for New York's multitudes, and that it will afford a direct avenue of approach to the scenic attractions of the Highlands, the cities on the west shore of the Hudson, the Catskill Mountains and the western part of the state, as well as a direct route for the largely increasing traffic between southern New England and the west side of the Hudson determined the selection of this site for the first highway bridge to be built across the Hudson River south of Albany. The Bear Mountain Hudson River Bridge Co. secured a charter from the legislature in March 1922, and having obtained the necessary rights-of-way and permits, and completed its financing, the contract for construction was executed March 24, 1923.[14]

Smith continued to list the various engineers and organizations involved in the bridge's building and design. He added that "the cable wires and suspender ropes are being manufactured by John A. Roebling's Sons Co. of New York"—the company that designed and built the iconic Brooklyn Bridge forty years earlier.[15] As the project moved forward with engineers on the ground laying out the course of work to be followed, public excitement continued to build. Area residents devoured all information they could find about the venture.

Apart from his roles managing the construction and public relations, Roland was still tying up loose ends concerning the financing arrangements. The final list of chief investors included E. Roland Harriman (of course) along with Mary W. Harriman, his mother; Dorothy P. Freeman, sister of George W. Perkins Jr.; and Evalina B. Perkins, mother of George W. Perkins Jr.[16] Correspondence between Harriman and the Newburgh Chamber of Commerce, as well as solicitations directed at Yale University, provide insight into W. A. Harriman Co.'s efforts to seek investor commitments from organizations and individuals. A letter from Joseph W. Wear, an employee in the Philadelphia office of W. A. Harriman Co., to Roland, dated April 3, 1923, indicates that Wear had reached out to Tom Farnam, the associate treasurer and comptroller of Yale, to inform him of the investment opportunity. However, Farnam responded to Wear that he would need the approval of Otto Bannard, chairman of the Yale Finance Committee, to secure Yale's support.[17] Wear suggested Roland speak to Bannard himself regarding the matter, since, as far as Wear understood, Roland had a relationship with the chairman.[18] It is unknown whether Yale decided to join the syndicate, but it is clear that Wear (and no doubt others in the W. A. Harriman organization) hoped the Bannard-Harriman relationship might positively influence the approval.

Regardless, the full collateral requirement of $1.5 million in income bonds along with the 12,500 shares of common stock was finally satisfied. On April 1, 1923, a mortgage and deed of trust for $3 million was entered into by BMHR Bridge Company and recorded with the Chase National Bank of the City of New York, serving as trustee. The document explained that the company, with the due consent of its holders of more than two-thirds of all outstanding capital stock, had determined to issue bonds in the aggregate principal amount not exceeding $3 million and to secure these bonds accordingly. The bonds, known as first-mortgage 7 percent, thirty-year sinking-fund gold bonds, were designated coupon bonds (simply, a bond upon which interest is paid by presenting a coupon) in the denominations of $100, $500, and $1,000. The bonds were dated April 1, 1923, payable in thirty years, on April 1, 1953. In the period in between, they would bear interest at a rate of 7 percent per annum, payable semiannually on the first day of October 1923 and then on the first day of April and October in each year thereafter until maturity. A sinking fund would be established to pay the principal amount of any bond prompted by purchase or redemption.[19]

With everything in order, the Bear Mountain Hudson River Bridge Company issued a notice announcing that the first annual stockholders meeting of the company would be held on Friday, April 20, 1923, at noon, at the office of the company, located at 39 Broadway, New York. The notice specified that the agenda would include a variety of specific matters to be discussed.[20]

The first meeting of the company did take place on April 20, 1923. If George W. Perkins Jr. was in town, we can only assume that as a major investor, he would have been in attendance. Records show that the work was started at the site of the Bear Mountain Bridge soon after the contracts were signed. In preparation for the start-up, surveys would have been done and geological studies ordered to ensure that the bridge was situated in the most strategic location.

18

Construction Begins

Frederick Tench was annoyed when he entered his office at the Bear Mountain Bridge construction site. Wilson Fitch Smith, the resident chief engineer representing the Bear Mountain Hudson River Bridge Company, knew it when Tench, the project's general contractor, pulled open the door and slammed it hard behind him. He stomped across the floor to where Smith was standing and launched into a tirade. McClintic Marshall Co., the steel fabricator and erector for the project (and future acquisitor of Bethlehem Steel), had just delayed the first delivery of tower steel yet again, growled Tench.

Bridge construction had begun in earnest on Tuesday, April 3, 1923, with the arrival of engineers and surveyors to lay out the western access road.[1] Since only one cofferdam would be installed as part of the construction of the four legs required for the two towers—the other three would sit on the native bedrock naturally present at the site—the builders projected that there would be no problem in having these footings completed and ready by the time the steel was scheduled to arrive in early August.[2] Enthusiasm was high and anticipation was building—the project was off and running!

But soon unforeseen "job conditions" made their presence known. Despite intense efforts on the part of the contractors to plan for every contingency, it became apparent that the project was not getting off to the start Terry & Tench Co., Inc., had envisioned. Besides the McClintic delays, progress at the excavation sites was being seriously held back by the endless amount of rock, which was extremely dense. The removal process was slow, creating the need for more equipment, more power, and more labor. The project had just started, costs were already increasing, and the schedule was not being met. It was up to Tench to determine what, if anything, could be done to remedy the problems and get the project back on track.

All this troubleshooting was happening in the background. As the contractors scrutinized the schedules and agonized over start dates behind closed doors, the Bear Mountain Bridge project began capturing the keen attention of the public. After all, the bridge promised to provide motorists immediate access to the opposite shore of the Hudson, expanding their world and their mobility. Articles about the bridge appeared in national and local papers, feeding a growing curiosity for details even among those individuals who opposed the bridge. In the July 18, 1923, issue of the *Outlook*, contributing editor Lawrence Abbott voiced his opinion that the Bear Mountain Bridge was designed without much, if any, regard for beauty, referring to it as a piece of "tin frumpery." Although Abbott acknowledged that he was not professionally competent to say what kind of bridge should be built at the location, he argued that he had enough aesthetic sense to provide his perspective.[3] This declaration stirred a pot of varying reactions. The *Engineering News-Record* responded with two editorials later in July, both of which challenged Abbott's comment that the bridge should "contribute to the beauty of the Hudson." The consensus, at the time, seemed to deny the idea that any bridge, no matter how beautiful, could add to the already consummate beauty of the river.[4]

On July 21, 1923, a very pointed rebuttal to Abbott's comments appeared in the *New York Times*, written by J. DuPratt White, president of the Palisades Interstate Park Commission, which was directly involved in the project and responsible for approving the bridge's design. White assured the reader that "the best architectural advice available had been sought in connection with the design of the automobile and toll bridge being built across the Hudson at Bear Mountain."[5] E. Roland Harriman, president of the Bear Mountain Hudson River Bridge Company, echoed White's comments. He was also quoted as saying that eminent engineering and architectural advice dictated that while stone towers (one of Abbott's recommendations) might have added to the bridge's attractiveness, the site's physical limitations prohibited them, as did the significant increase in cost they would precipitate. Ultimately, Harriman explained, it was decided that "it was best to provide a well-designed, adequate bridge for the benefit of the public rather than allow prohibitive costs to cause the project to be scrapped."[6] Years later, Howard Carter Baird discussed his reasoning for selecting the design in his dissertation on the Bear Mountain Hudson River Bridge and noted that "as the Manhattan Bridge, the Williamsburgh [sic] Bridge, and the Philadelphia Bridge (projected) were designed to have steel towers, it seemed entirely proper to adopt such construction, provided proper proportions be used and simplicity and unobtrusiveness be the keynote."[7]

In 1894, when Leffert Buck planned the design of the Williamsburg Bridge, the use of steel in bridges was relatively new. At the time critics complained about its clumsy appearance. Although Buck chose steel as his material, he clung to the old design of the previous period when wrought iron was the norm, which was clumsier and not as sleek as the "modern" steel designs. By the time Gustav Lindenthal and architect Henry Hornbostel designed the Manhattan Bridge (1903), the standard had evolved, and the Manhattan Bridge was seen as the product of new ideas and forward strides—no longer accenting utilitarian qualities but rather designed to please the eye. Lindenthal believed, "Our city will be preeminently the city of great bridges, representing emphatically for centuries to come the civilization of our age, the age of iron and steel."[8]

Otmar H. Amman, who designed the George Washington Bridge (1932), said, "It is a crime to build an ugly bridge." He believed, "Economics and utility are not the engineer's only concerns. He must temper his practicality with aesthetic sensitivity. His structures should please the eye. In fact, an engineer designing a bridge is justified in making a more expensive bridge for beauty's sake alone."[9]

Despite Mr. Abbott's early criticisms claiming frumpery, since its completion the Bear Mountain Bridge has been admired and lauded in many countries, including our own. Several articles of appreciation appeared in newspapers here and worldwide, testifying to its beauty and comparing the Hudson Valley to the renowned river valleys throughout Europe.

It seems that the warnings expressed by Abbott in those early months were overruled many times in local, national, and international newspapers.

Despite the back-and-forth criticisms being splashed across the papers, the project continued to move forward, and the people in charge attempted to meet the challenges as they were presented. Besides getting approval from the general contractor, Terry & Tench Co., Inc., all shop drawings had to be reviewed and approved by the PIPC engineers. The state charter had delegated the PIPC to act on behalf of the state of New York, vesting it with the authority to oversee the bridge's construction and to ensure that all state regulations were being met. Much of this responsibility fell to the chief engineer, William A. Welch, the PIPC's liaison to the project. In addition, Howard Carter Baird, the engineer responsible for the original design, would be on site to confirm that the work was being done in compliance with his drawings.

Progress reports were prepared and submitted monthly from the beginning of the project by Wilson Fitch Smith, providing clear, continuous communication between the field office and Roland Harriman as the project advanced. By May, while actual tower fabrication had not yet started, there had been some rivet-hole punching done on the lower plates. The monthly report, prepared by Smith, also confirmed that Tench had secured a promise from McClintic Marshall for tower fabrication to begin no later than mid-July, resulting in the first delivery of tower steel to the site by August 10. (McClintic also confirmed that they had received approximately 95 percent of the raw material needed for the tower steel and 70 percent of the raw material for the approach spans.) Not everything was back on track, however. The May report also mentioned that there would be a delay of at least six weeks on the delivery of the anchor bars, a fact that hadn't changed by the time the June report was issued.[10] The news about the suspender ropes and cable wires was more promising when it came. The contract had been awarded to John A. Roebling's Sons Co., operating out of Trenton, New Jersey. The preliminary processes for fabricating the cabling, both the main catenaries and the deck suspenders, were well under way by the end of June.

Progress at the bridge site during June included site-preparation work. The Western Union telegraph wires along the railroad tracks were successfully removed and relocated, and the New York Central Railroad installed turnouts and tracks for construction sidings on both sides of the river. And, critically, a power plant was positioned on the east side of the river to provide electricity and compressed air necessary to operate the equipment and tools used in the excavation and construction. On both sides of the river, work began on the anchorage tunnels.

Excavation for the east-approach highway also began, but as mentioned earlier, the presence of almost all pure rock caused the process to move much more slowly than expected. Tench ordered an additional air compressor and a second steam shovel, along with the necessary crew to operate them, to ensure the required schedule was maintained. The situation of the new steam shovel was a major accomplishment in and of itself. The lumbering, awkward machine, which weighed about twenty-nine tons, could not be moved into position as easily as first thought. The idea was floated to disassemble it and haul the components up the side of Anthony's Nose; fortunately, logic prevailed, and that plan was discarded in favor of bringing it onto the site from the existing road just north of the area, allowing it to be driven forward while mounted on a caterpillar tractor. This method was successful, and as it climbed to its perch six hundred feet above the New York Central tracks, all breathed a sigh of relief.[11]

In mid-July Wilson Fitch Smith and Howard Carter Baird visited the McClintic Marshall plants in Pottstown and Birdsboro, Pennsylvania, where the project's steel components were being fabricated. Because of reported delays, the men decided to personally monitor the subcontractor's scheduling and assess the progress being made. Smith confirmed in a letter to E. Roland Harriman, dated July 14, 1923, that several issues were to blame for the delays. He explained that the first steel delivery out of the Pottstown plant was being pushed back two weeks from McClintic's initial promise to Tench because of a shortage of competent labor. "The lack of activity in the shops would bear this out, but other work apparently has something to do with it," wrote Smith, offering his assessment as to why the fabrication was behind schedule.[12] Apparently, McClintic not only had failed to hire and train adequate staff at their Pottstown location, but was also moving another project out of sequence and putting it ahead of the BMHR Bridge project. A visit to the Birdsboro foundry was no less discouraging. Here, Smith and Baird learned that McClintic had allowed the process of pouring the base castings to be interrupted and restaged, again resulting in delays.

As part of their contract, McClintic had agreed to cast four end sections of the base castings at intervals of about six days, rebuild the pattern, cast the middle sections, and then go back to the remaining four end sections for the final casting. Each casting required seven days to cool, seven days to anneal, and six days to machine—in total nearly three weeks for each casting from the date of pouring. The castings would then have to be assembled in the shop, adding more time before shipment could be scheduled. During their visit to the plants, Smith and Baird secured assurances from McClintic that this original plan would be adhered to from that point forward. Based on the agreement, they were assured that one complete set of base castings would be ready to ship by August 9, one set by August 28, another by September 17, and the final set by September 28. The erection of the first tower depended on the delivery of fabricated material from McClintic's Pottstown plant, and the erection of the second tower would be determined by the delivery of the base castings from Birdsboro. Per Smith's July 14 letter, this timetable would translate into an expected delay from the original schedule of two months on the first tower and three months on the second. At the same time, delays in the field indicated that the anchorages would not be ready until roughly the same date. These combined conditions would result in a scheduling delay of three months, with little hope of any possible adjustments.[13]

The July job report confirmed that McClintic Marshall had pushed back the initial shipment of tower steel (promised for August 10) to August 24, with the final shipment to arrive by October 10, 1923. Fabrication was under way on the tower sections by July 15. Birdsboro communicated to Tench that the base castings were also in fabrication and confirmed that the first shipment would arrive on August 10 and the remainder on September 25, 1923, per the mid-July meeting. In the field, the concrete pour was proceeding apace. The northwest tower foundation was completed on August 1 and the northeast foundation on August 8.

The unanticipated amount of rock present in the area continued to slow the excavation within the cofferdam for the southwest foundation.

However, it was expected to be ready for concrete by the end of August. Excavation for the anchorages on the west side was held up for lack of compressed air early in July, but with the installation of a new compressor by the end of the month satisfactory progress resumed from that point forward. The east-anchorage excavation started in late July. The east side provided the best condition, as the anchorage would be drilled directly into the mountain of Anthony's Nose. The west anchorage required longer tunnels owing to the existing topography, but both locations were ideal from an engineering perspective. Once complete, they would provide extreme holding power to counter the massive pull of the cables.[14] Progress was consistent but slow because of the dense rock.

Work on the east-approach highway toward Peekskill continued to experience setbacks. Although rough grading had been completed for a length of 1,310 feet, the final layout for the eastern highway could not be confirmed until approval was received for a right-of-way. Franklin Couch and his wife, Mary, and Clifford Couch and his wife, Carolyn, together owned a tract of land in the vicinity of Roa Hook Road, in the town of Cortlandt, and the parcel appeared to be in the direct line of the new eastern-approach highway around Anthony's Nose. Before the project's commencement, a discussion between the Palisades Interstate Park Commission and the owners had yielded an agreement that the Couches would grant a right-of-way so the approach highway could pass through. As the project progressed, however, it seemed there were difficulties in getting the legal document prepared and signed off. The BMHR Bridge Company, along with the PIPC, was hopeful that this contract would be secured no later than September 1, 1923.

The original schedule had projected that by August 1, construction would be 100 percent complete on the tower foundations, piers and abutments, anchorage excavations, and western-approach highway grading. The schedule also projected that the erection of the towers would be at least 33 percent complete by that time and that there would be reasonable progress on the eastern-approach highway. But according to the job report dated July 31, 1923, these milestones had not yet been accomplished, and the overall project was woefully behind schedule. That same report confirmed the expected delays for the steel deliveries that were explained in

Wilson Fitch Smith's July 14 letter to Roland Harriman. The delayed steel deliveries would create a three-month lag on the erection of the towers and, ultimately, the overall project completion.[15]

Additionally, around the time Smith sent his letter, there appeared to be some conflict between Frederick Tench and Roland Harriman regarding contractor billing. According to correspondence between the two dated July 6, Tench maintained that it was his understanding, per the construction agreement of March 24, 1923, that certain preliminary expenses (such as site preparation) were to be advanced to his firm according to subdivision 2 of Article XI, and that the 15 percent retainage outlined in the contract was to exclude material costs. Tench's position was that he should be paid in full for materials he had already paid for but may not yet have arrived on the site or been installed. He provided documentation from McClintic's order department and confirmed that McClintic was forwarding copies of all the steel-mill shipping invoices to him (he would then send them on to Harriman as proof of payment).[16] The inference we might make here is that perhaps the contractor was experiencing some financial problems, and without the cooperation of BMHR Bridge Company, he was going to have difficulty maintaining the project. Harriman did not agree with Tench's interpretation of the contract and responded that, while the bridge company wanted to cooperate fully, the two parties should refer the matter of legal interpretation of the contract to their respective attorneys. Additionally, he suggested that Tench consult the bonding company for any financial assistance he may need. If there should be any problems with this course of action that might result in a delay or interruption of the work, he wrote, the bridge company would consider providing financial assistance, but only if Terry & Tench Co., Inc., disclosed their full financial position for the bridge company's review.[17]

By the end of August, the site was still awaiting the first shipment of steel. Despite several updated delivery dates, the most recent of which had been August 24, McClintic Marshall continued to be delinquent. Smith continued to track McClintic's progress. His August project report to Harriman indicated that while 60 percent of the raw material, or 6,013 tons, had been received by now at the McClintic plant, only a small portion had been fabricated. Although the fabrication was moving forward, progress was much too slow. Both Tench and Smith continued to monitor the situation with growing frustration. By the end of August, not only had no steel found its way to the site, but no alternative date for its arrival had been confirmed. Things were going better at the foundry in Birdsboro, which had cast all the tower-base castings by this time. Shipments to the site were being scheduled, the first of which was received on August 31. McClintic reported that the American Bridge Company had received all the material necessary for the eyebars and that approximately 12 percent of the project requirements had been completed. The Roebling Company reported that fabrication of the suspension-cable wire as well as the hanger cables was well under way.

At the construction site, some progress was being made, but the estimated delays remained a frustrating reality for Smith and Baird. Anchorage excavation was moving slowly; the western tunnel was approximately 50 percent complete, and the eastern tunnel had not progressed according to expectations because of a lack of skilled labor, excessively hard rock, and difficult site access. The southeast tower-foundation concrete was in place by September 4, and the cofferdam at the southwest tower foundation was expected to be ready for concrete by mid-September. One tower-base casting was received on site by the end of August and was scheduled to be set by mid-September. Excavation for the retaining-wall foundation on the north side of the eastern plaza had started but was also moving very slowly. The rough grading for the eastern-approach highway was active, but weather conditions continued to affect its progress, which was not anywhere near what had been originally promised. At the end of August and the beginning of September, work on the tower foundations had progressed to about 73 percent, anchorage excavations to 40 percent, and eastern-approach highway gradings to 18 percent. Still, no work had been initiated on the piers and abutments, the tower erection, approach spans, or western-approach highway grading. The three-month projected delay remained unchanged.[18]

On the first of October, Smith reported to Harriman that McClintic had received almost all the material required for tower fabrication. The specification was for silicon steel, a material considered to be reliable since it was used in the Memphis Bridge and the New London Bridge, designed by Ralph Modjeski. Although nickel steel was used in the Manhattan Bridge, it was deemed too expensive for this project. One tower leg had been assembled in the shop up to the second splice, and the other three had been assembled to the first splice, but the first shipment of tower steel would not be delivered to the site before mid-October. The towers were designed to consider not only appearance but also ease of erection and accessibility for inspection and maintenance, with bracing being spaced to allow for a man to enter the openings. The structural integrity was reinforced by the simplicity and uniformity of the members' arrangement and, at the same time, resulted in a smooth, clean appearance.[19]

All the base castings for the towers had arrived, and two of them were in place by October 1. Smith noted in his October report that all the steel castings and all the eyebars for the anchorages had been fabricated, with the majority of them being received at the site and the balance scheduled to follow soon. Roebling reported to the BMHR Bridge Company that 80 percent of the suspenders had been manufactured by October 1. The wires were being spun at Roebling's Trenton facility at the time the report was issued, and delivery to the site was promised shortly. Both east and west anchorages were now approximately 60 percent complete. Tench explained that they would not need to be complete until the towers were set, which, at this point, was not expected to occur before January 1, 1924.[20]

The long list of issues plaguing the project was causing multiple delays. Tench fielded the problems and handled the stress as he juggled the scheduling and attempted to counter as many delays as possible. The sooner the balance of the steel castings and the anchorage eyebars were installed, the sooner he would receive his payment. A progress payment was typically billed according to the amount of work completed each month. When the work was delayed, payments were withheld—which was exactly what was happening.

Excavation for the piers on the western-approach span started in September and was reported to be about 70 percent complete by October 1, with plans to place the concrete on all six piers before the end of the month. In early September drilling for the east abutment had begun, as well as for the eastern plaza retaining-wall footing; both items were key to moving forward on the anchorage tunnels and approach-highway grading.

Work had been initiated at the southwest pier as well. The excavation of its cofferdam was completed, and the first concrete was placed on September 21. The base castings at the east-tower foundation were placed between September 22 and 24, but still had to be grouted. Progress on the eastern-approach highway's rough grading slowed during September, with very little work completed on the steep slope overlooking the New York Central Railroad tracks.

Around this time Terry & Tench furnished a revised schedule that put the bridge's completion before December 1, 1924. However, the progress of the work did not align with that prediction, nor was there evidence that the urgency of the remaining tasks was fully understood or that necessary steps were being taken to meet the scheduled dates. Not to mention, there was still no word on the Couch right-of-way, which was necessary to finalize the layout of the eastern-approach roadway.[21] However, the legal paperwork for the acquisition of some other parcels of land located on the west shore had been fully executed as of September 25, 1923. Two of the parcels were currently owned by the state of New York, so there was no issue with ownership or rights-of-way for them, but there were other problems: one parcel had been gifted to the state by Rowland Hazard and his wife, Helen Campbell, in April 1916, with the caveat that it was to be used as parkland only. Another had been donated by the Harriman family in 1909, with the same stipulation. Both parcels included the qualification that if, at some future date, the land was ever used for something other than a park, the parcels would revert to the donor or their heirs. To resolve this potential sticking point, in both instances, representatives of the donor families signed documents indicating their acceptance of the bridge as part of the park system so that the project could move forward unencumbered.[22] On the eastern shore, however, the Couch right-of-way problem remained, with no communication from the owners as to how they wanted to proceed. If this issue wasn't resolved soon, it would become a major scheduling headache for Smith and Tench.

On November 1, 1923, the first section of the east-tower steel finally arrived on site along with all the eyebars for the anchorages. Roebling had most of the cable wire ready but was forced to suspend operations while they waited for the delay to be resolved. The tower foundations were completed, but, at this point, construction of the towers was still about four months behind schedule. Among other issues, there was an accident involving a derrick being used on site in the excavation of the east anchorage. The mishap almost resulted in a complete shutdown of work during October. And while the east abutment had been prepped for blasting, the blasting had not yet been accomplished. Progress on the east-approach highway was still extremely slow in October, with only about 28 percent of the grading completed to date. Overall, October's work pace was disappointing and not at all in line with scheduling requirements.

Still, there was some success reported in the November 1 report. The west-pier excavation was completed, and concrete was being poured, and, as noted, minor progress had been made on the towers. But perhaps the best news was that the paperwork for the Couch property was finally in the hands of counsel.[23]

By early December 1923, McClintic had assembled 75 percent of the towers in their shop, and 30 percent had shipped to the site; 100 percent of the shore spans were also assembled, with 22 percent shipped to the site. The stiffening trusses had not yet been started, but most of the material for them had been received in-house at McClintic's plant. One hundred percent of the anchor bars were completed and were shipped, along with 82 percent of the steel castings. By this time, Roebling had also finished all suspender ropes and was slowly releasing them for shipment to the site. Excavation on the east anchorage was now 62 percent complete and on the west 85 percent complete. Both the east and the west towers were in the process of being erected; the east tower would go up first, with the west tower following approximately one month later.[24]

Despite these small advances, the overall project schedule had not progressed at all over the previous month's projections, and the pace was reported as unsatisfactory. Notably, the east abutment was dangerously behind schedule owing to the contractor's inability to provide adequate compressed air, a recurring problem; it was needed to power the hydraulic jackhammers and drills used to break up the rock. On the bright side, the long-awaited resolution to the Couch property was finally achieved.[25] A negotiation had taken place between the landowners and the PIPC in conjunction with the BMHR Bridge Company that resulted in the Couches selling the parcel outright to the state rather than granting a right-of-way. Once that agreement was reached, the documents were prepared and the parties signed off.[26] The project report dated December 1, 1923, indicated that a contract for the sale of the Couch property had been executed, and the work in that area could now proceed without issue.[27]

As to the steel contract, McClintic finally accelerated the shipments of assembled material to the site during December. One hundred percent of the material needed for the towers, shore spans, and anchor bars had been received at McClintic's plant, and most of these assemblies had been constructed and shipped. Stiffening trusses and floor systems were in fabrication. The material needed to maintain site progress and not delay the project further had been delivered and was available to the erection crews.

Excavation of the east and west anchorages, however, fell short of the December 1 deadline promised by Terry & Tench in November, and there was little progress during the month. Per the project report dated January 2, 1924, the excavation was five months behind contract schedule, nearing only 82 percent completion. The east-tower completion date had been pushed out to February 1, 1924, a full four months later than the original date, while the west-tower completion date was now projected as March 1, 1924—a delay of five months. In addition, the excavation and grading of the eastern-approach highway continued to be about four months behind.[28]

The same report listed the completion date of the west-tunnel excavation as January 7.[29] Excavation on the east abutment had progressed very little, but the north retaining wall was built within ten feet of the top. A new boiler was installed at the east-tower powerhouse in late December, replacing one that had failed. This new equipment would provide necessary power to continue the construction.

The east tower was erected through the fourth section by the end of December, and a riveting gang began their follow-up work on December 26. Riveting was considered an extremely dangerous job, again owing to the heights and the awkward access to the areas to be riveted. Since the late nineteenth century, members of the Mohawk nation had been acknowledged as the most sought-after riveters in the New York area. Members of the Kahnawake Reservation near Montreal, Canada, first started in the trade when the Victoria Bridge across the St. Lawrence River was being built on a parcel of the reservation's land. The "Skywalker" tradition has continued through generations. Indigenous riveting gangs helped to build the Chrysler Building, Empire State Building, and Rockefeller Plaza as well as the George Washington Bridge, the Bayonne Bridge, the Triborough Bridge, the Bronx-Whitestone, and many others. The Mohawks had great proficiency for the task and demonstrated no fear of heights. In the early 1900s, they worked on bridges from Canada south to New York and Pennsylvania and were members of the steelworkers' unions, specifically the Brooklyn Local 361. By the 1920s large numbers moved to the New York City marketplace, filling jobs that were plentiful owing to a building boom that continued as the Depression-era public works projects appeared. Although ironworker technology has improved over the years, there are still between thirty-five to fifty fatalities per year, mostly from falls. Whether any of the riveters on the Bear Mountain Bridge were from the Mohawk reservation, we cannot be certain, but the ties between the Mohawk nation and the steel trade in the New York metropolitan area are more than six generations strong.[30]

The first section of the west tower was erected in early December, and a creeper and derrick arrived by the twenty-first, enabling the crew to move forward with the second section, which was in place and almost completely installed as of the date the report was issued (January 2). The *Engineering News-Record* explained the novel way in which the towers were being erected: a stiff-leg derrick was used, which was the same process used to erect the Manhattan Bridge spanning the East River, in 1909. The process involved deploying a creeper traveler (which is a platform that rides on wheels set into rails that run along the structure on which it is anchored). The creeper carried a fifty-ton capacity stiff-leg derrick on the river side of the tower. (A stiff-leg derrick is a type of crane used to lift exceptionally heavy pieces of equipment and stabilize them while they are being anchored into place.) As each section of the tower was lifted and incorporated into the structure, a hoisting engine pulled the creeper up to the next level from which additional tower sections would then be lifted and set into place. The creeper traveler used had been specifically conceived by Terry & Tench to accommodate the unique characteristics of the two tower posts, which were designed to come closer as they rose toward the top of the structure. The heaviest section to be lifted weighed fifty tons and was fifty-five feet long.[31] The erection of the towers was done over six months from December to May, involving the placement of four thousand tons of steel framing while fighting extremely cold weather and high winds that often shut down operations. The icing of the river frequently interfered with transporting men and materials to the tower locations.[32]

The east-approach highway excavation and rough grading did not progress much in December because of another continuing issue—problems with air compressors. Four new gasoline compressors were finally installed at the end of the month, which were more than enough to support the remaining excavation work.

Concerns about the completion date had amplified in December, with that month's report now projecting the "ready for traffic" date as April 1, 1925, and the full completion date as June 1, 1925,[33] still a full five months behind the original completion date of January 1, 1925, promised by Terry & Tench in their original estimate to the BMHR Bridge Company. Although it was the reality, the myriad issues and delays were unknown to the public, and the local papers continued to jubilantly track the bridge's progress. In early January 1924, photographs appeared on the front page of the *Peekskill (NY) Evening Star*, announcing to residents that the east tower was now 156 feet tall and the west tower 70 feet tall. As far as the residents were concerned, progress was indeed being made.[34]

Job reports from early February indicate that McClintic was finally keeping up with the steel fabrication. The east tower was erected through the seventh tier, the west tower through the fifth tier, and riveting was moving favorably on both.

The Bear Mountain Bridge employed a variety of methods used in several bridges of the time. Besides the use of a creeper traveler with a stiff-leg derrick, explained earlier, the Bear Mountain Bridge and the Manhattan Bridge were both designed as suspension bridges utilizing parallel wire-cable fabrication—manufactured and installed by John A. Roebling's Sons Company. Before the Manhattan Bridge, the Williamsburg Bridge across the East River was also of parallel wire-cable construction. The Bear

Mountain Bridge shares several statistics with the Williamsburg. The Williamsburg towers are 350 feet high, and the Bear Mountain Bridge towers are 350 feet high; the main span of the Williamsburg is 1,600 feet, and the main span of the Bear Mountain Bridge is 1,632 feet; both bridges were pioneers in using reinforced concrete decks; and both bridges utilized galvanized-steel cable wire. Leffert Buck, the engineer-designer of the Manhattan Bridge, opted to use nongalvanized-steel cable wire, and this small detail resulted in extensive corrosion problems during the period following that bridge's construction.[35]

The project at Anthony's Nose continued to struggle with material delays and consistent difficulties in drilling the unprecedented amount of solid granite present throughout the site. Although excavation on both the east and the west anchorages was finished by February, very little progress had been made on the east abutment and plaza. Construction issues weren't the only obstacles slowing things down. In early February, Terry & Tench's representatives made several field-personnel changes, creating an atmosphere of uncertainty, resulting in low morale among the workmen. Poor weather conditions throughout January added more stress and additional delays. Management took advantage of a delay-induced lull during this period to scrutinize the plans and methods being used and to look for ways to improve production and morale.[36] Despite Terry & Tench's attempts to fight the setbacks, payments were delayed, further impacting the company's finances.

Something had to be done to reset the schedule—specifically relating to the eastern-approach highway. In early February a decision was made. Terry & Tench Co., Inc., would hire the firm of W. F. Carey and Company, Inc., a road-construction specialist, to take over this work. Also, a new project supervisor was hired to oversee the entire project on behalf of Terry & Tench Co., Inc., and Globe Insurance.[37] J. V. W. Reynders was well known throughout the steel industry, having managed the Pennsylvania Steel Company. He had been successful in building many bridges across the United States and abroad. By February 15 Reynders had already reached out to George W. Kittredge, the chief engineer for the New York Central Railroad, without whose cooperation and coordination the east-approach road could not move forward.[38] The reorganization, coupled with the arrival of Reynders's and W. F. Carey's additional crews, appeared to have resulted in a considerable improvement to both the schedule and the morale of the men. In line with these changes, the workforce was increased by about fifty men at the bridge and seventy men on the roadway.[39]

By early March, it had become clear that progress was being made, despite unfavorable weather conditions. For their part, both McClintic Marshall and Roebling were also moving forward. According to the job report dated March 1, both towers were complete except for the top transverse bracing and riveting, and these elements were in progress.[40]

Reynders may have decided to concentrate on the excavation for the east abutment because the delays in this area had begun to interfere with the installation of the east-approach span. At the same time, the progress on the excavation for the highway was improving and preparations to move northward were being made so that the crews could start working once the frost was off the ground. The March report also indicated that the tunnels would be ready to accommodate cabling for the temporary footbridges (also referred to as "foot walks") by April 1. The footbridges were working platforms, located immediately beneath the two main cables, with the footbridge cables anchored to girders installed in the anchorage pits.[41] These narrow platforms or walkways were essential for the bridgemen to work from while they stretched the main cables, the suspension bridge's primary support. Once the main cabling was complete, these platforms would no longer be necessary so they would be removed, and the footbridge cables would be dropped and reconnected to the bridge structure as suspenders.

March arrived and with it the advent of spring. The metaphor of rebirth could have hardly been more appropriate. The project had finally reached a turning point, reflected in the men's newly positive morale and, most important, the amount of work they were accomplishing.[42]

The job report dated April 1 indicated that good progress had been made on both the steel erection and the cabling during the previous month. The report specified that the stringing of the cables for the footbridges would begin on April 15. Excavation for the east abutment was complete, and excavation for the plaza was very close to being complete; as well, riveting of the east tower was 61 percent done, and the west-tower riveting was 90 percent done. Progress was also being made on both the east- and the west-approach spans. Once these roadways were complete, they would become staging areas during the erection of the main suspension span.[43]

Around this time, it was determined that the area known as the "Couch swamp"—named for the previous landowners—required a good amount of fill. The process got under way quickly, and by late March approximately 20 percent of the required fill was in place. There was discussion among the engineers along with Reynders and Smith about the feasibility of building a viaduct rather than cutting and filling the northerly twenty-five hundred feet of roadway near the New York Central Railroad tracks, but early in April the decision was still pending. Alleged advantages included possible savings of time and money; critics of the plan pointed out that the viaduct would be a less permanent structure, potentially requiring increased maintenance, resulting in additional costs in the long run. The roadway would be narrower and subject to the ever-present danger of rockslides—potential issues that could not be ignored.[44] The final decision would come later.

Even with the recent progress and improved planning, the project was still behind schedule, and the team of Reynders, Smith, Tench, and Baird continued to seek ways to speed things up. To that end, in April officials from Terry & Tench Co., Inc., and the New York Central Railroad met to resolve an issue that had developed on the east-approach highway, immediately above the New York Central Railroad tracks. After the required blasting and chiseling of granite at Anthony's Nose was completed, an overhanging ledge remained that was considered a menace. The officials agreed that it should be removed, and a new round of blasting was scheduled for 11:30 a.m. on April 11, 1924. As representatives of both entities looked on and with the scheduled train run paused, an explosive charge was detonated. The ledge, composed of tons of rock, fell to the tracks below, damaging the track, several railroad ties, and various areas of the roadbed. While the railroad had announced a temporary suspension of service that day, it hadn't informed passengers about the blasting, and according to local newspaper reports, passengers were left to wonder what was causing their trains to be delayed several hours. In the meantime, the workmen removed the debris and replaced the twisted railroad ties, which was accomplished by four o'clock in the afternoon, when railroad operations finally returned to normal.[45]

By early May, most of McClintic Marshall's contract work was nearly complete. All that was left were the stiffening trusses, deck systems, and steel castings, all of which were being worked on. The west anchorage was complete, and the east anchorage was only slightly behind, as the castings were being placed.[46]

As scheduled, the footbridge cables were strung in April, with the expectation that the walkways would be finished by the end of May. Indeed, on May 21, 1924, the *Peekskill (NY) Evening Star* reported that "the bridge was completed and formed the first real connecting link across the historic Hudson south of Poughkeepsie." The reference, of course, was not to the formal bridge structure, but rather to the completion of two footbridges that would allow the workmen access to the full length of the bridge to complete the work; each footbridge consisted of wooden slats supported by steel cables. "For the first time in history, engineers . . . were able to walk across the Hudson River on two suspension footbridges," reported the paper. "The two footbridges [were] supported by four cables each stretched from a high steel tower located on the Peekskill side of the river and connected with another such tower located near Bear Mountain Park."[47] These cables were the support for the walkways that would be erected directly below and about three feet from the final contour of each of the proposed main bridge cables. The footbridges would vary between six and fourteen feet wide depending on the size of the main cables. They must be wide enough to allow for the spinning of two strands on each side of the proposed cable, and, of course, the workmen must have enough room to move freely about and be able to access the cable throughout the spinning process. Trussed cross bridges would be installed at mid- and quarter-points of the spans to stiffen the footbridges laterally, along with a system of storm cables suspended below the footbridges. These storm cables would be anchored to the tower bases and connected to the footbridges at several different points with guy wires. Towers erected on the footbridges supported the aerial rope tramway and were spaced appropriately to allow efficient tramway operation. Typically, the footbridge ropes were made to meet the specifications of the suspender ropes for the finished bridge so that once the men completed the spinning and compacting of the main cable, they would remove the footbridge ropes and cut them into the necessary lengths, transforming them into suspender ropes.

The maximum stresses in the footbridges were calculated so that the required lengths of footbridge rope would be installed to comply with the required safety factor. For each footbridge, it would be customary to use two groups of two, three, or five lengths each. If the number of ropes for each group was more than two, they would be placed in two layers and clamped together at the floor-beam points, thus maintaining a uniform cable throughout its entire length. These ropes would then be adjusted so that the footbridges hung in correct relation to the proposed cables.[48]

The framework of the footbridges consisted of floor beams, stringers, and braces covered by floorboards with cleats placed to allow for proper footing on steep grades; all components were of timber construction, and because of an innovation devised by Riley Coppage, the superintendent in charge of cable erection, the components were framed on the ground and then lifted to the top of the towers. This change in process cut the time required for installation considerably—a factor sorely needed at this job site. Components on both sides of each tower were fabricated and installed at the same time and attached in sections. To maintain balanced stresses at the towers, the work of erection would have been carried out on each span simultaneously.[49]

While the footbridge work was being performed, the cable-spinning machinery was installed so that once the footbridges were complete, cable spinning for the main cables could start, as scheduled, on the first of June.

It seemed that things were finally coming together. Both towers were 99 percent complete, the east-approach span was 96 percent complete, the west-approach span was 94 percent, and the fill for the Couch swamp on the east-approach highway was 75 percent installed. The proposal to incorporate a viaduct for approximately one thousand feet on the northern part of the highway had been reconsidered, and the decision was made instead to cut and fill the area in question and create a permanent roadway. (This change in contract was reviewed and approved by Major William A. Welch, consulting engineer for the PIPC.) Reports indicated that April was 39 percent more productive than March, even though payroll costs rose only 12 percent.[50]

By June 1, 20 percent of McClintic's contract remained, with only the stiffening trusses yet to be delivered. Roebling had produced at least 70 percent of the required cable wire. Roebling's footbridges were complete, and the main cable spinning was ready to begin. East and west anchorages were finally done, and both approach spans were also nearing completion. The Couch swamp had been filled.[51]

14. Plate 2, excavation of the western anchorage pits. Courtesy of the Roebling Museum; *Construction of Parallel Wire Cables for Suspension Bridges.*

15. Plate 4, tower nearing completion. Courtesy of the Roebling Museum; *Construction of Parallel Wire Cables for Suspension Bridges.*

16. Plate 7, construction of the footbridges. Courtesy of the Roebling Museum; *Construction of Parallel Wire Cables for Suspension Bridges.*

17. Plate 10, men spinning cables. Courtesy of the Roebling Museum; *Construction of Parallel Wire Cables for Suspension Bridges.*

18. Plate 15, adjusting cables in anchorage. Courtesy of the Roebling Museum; *Construction of Parallel Wire Cables for Suspension Bridges.*

19.1. Plate 21, looking east, wheels carrying cables. Courtesy of the Roebling Museum; *Construction of Parallel Wire Cables for Suspension Bridges.*
19.2. Plate 22, looking east, wheels carrying cables. Courtesy of the Roebling Museum; *Construction of Parallel Wire Cables for Suspension Bridges.*

20. Plate 32, road construction of the east-approach highway. Courtesy of the Roebling Museum; *Construction of Parallel Wire Cables for Suspension Bridges.*

21. Plate 37, squeezing gang compacting cable. Courtesy of the Roebling Museum; *Construction of Parallel Wire Cables for Suspension Bridges.*

22. Plate 43, western anchorage. Courtesy of the Roebling Museum; *Construction of Parallel Wire Cables for Suspension Bridges.*

23. Plate 49, suspenders in place, trusses under way. Courtesy of the Roebling Museum; *Construction of Parallel Wire Cables for Suspension Bridges.*

24. Plate 51, placing the final truss connecting the east and west portions of the bridge. Courtesy of the Roebling Museum; *Construction of Parallel Wire Cables for Suspension Bridges.*

25. Plate 53, correct cable construction. Courtesy of the Roebling Museum; *Construction of Parallel Wire Cables for Suspension Bridges*.

26.1. Plate 63, handrail ropes, east anchorage. Courtesy of the Roebling Museum; *Construction of Parallel Wire Cables for Suspension Bridges.*
26.2. Plate 64, handrail ropes, east anchorage. Courtesy of the Roebling Museum; *Construction of Parallel Wire Cables for Suspension Bridges.*

27. Plate 67, commemorative plaque on the bridge on opening day. Courtesy of the Roebling Museum; *Construction of Parallel Wire Cables for Suspension Bridges.*

28. Plate 68, the finished bridge. Courtesy of the Roebling Museum; *Construction of Parallel Wire Cables for Suspension Bridges.*

19

Innovations

The Bear Mountain Bridge was a turning point in the history of parallel wire-cable fabrication since several innovative processes were introduced by the Roebling company during its construction. Some of these breakthroughs involved processes that enabled the cable spinners to produce practically any diameter cable with any number of wires. While Roebling pioneered the use of steel as the main component of cable wire in the Brooklyn Bridge and first coated the wire with zinc (galvanizing) in that same project, both these conditions were still considered groundbreaking when they were used in the construction of the Bear Mountain Bridge.[1]

Specifically, a completely new process involving a flat-wire stay used in place of the round-wire stays previously considered the standard was developed during the construction at Bear Mountain. This innovation sped up the squeezing of the main cables. In the original process, multiple strands were bound together at intervals with round serving wires, forming the core of the main cable. All the round serving wires were then removed, and a predetermined number of strands, bound using a similar process, were placed around the core, creating an outer layer and several successive layers around the outer one, until the cable was complete. Squeezing jacks would then compact the cable to its final diameter. As each bundle of wires bound with serving wires was readied for compaction, the serving wires had to be removed quickly, at the last minute (with much dexterity by the crew member assigned this task, lest a finger remain as the final squeezing took place). It was a slow and tedious process. The new method perfected during the construction of the Bear Mountain Bridge consisted of a similar sequence except that a flat-wire serving band was used in place of the round, and because of its flatness it did not have to be removed at each progressive step. The flat bands were left in place on the inner layers and were removed only from the outer strands, giving the cable a smooth, monolithic appearance. This shortened the compaction time and thus the overall cable-spinning time, not to mention the safety aspect.[2]

Although the scheduling issues appeared to be improving, the financial concerns caused by the numerous delays Terry & Tench endured seemed to be beyond correction, and the problems that were created continued to play out behind the scenes. Correspondence between E. R. Harriman and J. V. W. Reynders in February 1924, shortly after Reynders came on board, indicates that Reynders was taking charge of the project and was preparing a full schedule of expected progress, that is, dates specifying when certain milestones would be reached, along with a schedule of expected financial disbursements to be released as those milestones were accomplished. Harriman seemed to be in sync with Reynders and expressed the BMHR Bridge Company's willingness to extend financial support according to those schedules. Reynders said he was comfortable with the projected schedule, and the only area of uncertainty had to do with the eastern-approach road and the cooperation of the New York Central Railroad, which would impact the road schedule greatly. It is uncertain how this scheduling process played out since there was no follow-up correspondence available, but if the process had been completely successful, there would have been no need to activate the performance bond. Also, there is no way of knowing where the accountability for failure should have been laid. Regardless, four months later, in June 1924, the BMHR Bridge Company called in the performance bond. Because of insufficient documentation, we cannot determine what the exact details were and just how the whole situation played out. Whether the hiring of W. F. Carey and Co., Inc., in February was Terry & Tench's attempt to reset the momentum of the project and prevent the calling of the bond, or whether things were already in such a downward spiral the BMHR Bridge Company forced them to bring in help to ensure the progress of the bridge construction, is left to the reader's speculation. What we do know is that the earlier concerns about the contractor's financial difficulties proved to be warranted. For Frederick Tench, though, it may have all seemed a bit anticlimactic. After all his efforts to get the Bear Mountain Bridge built, he did not reap many benefits. The construction of the Goat's Trail (the nickname bestowed on the eastern-approach highway) may have proved to be his Waterloo. The unforeseen costs of chiseling by hand through the solid granite that was the core of Anthony's Nose, coupled with the difficulties involved in building a road hovering for the length of a mile over one of the busiest railroads in the region—all while the railroad continued to run—proved to be a challenge that nearly bested him. The construction of the highway had to be coordinated with the New York Central Railroad's active schedule during every step of the entire project, with the priority, of course, being the safety and timely operation of the rail—not Terry & Tench's bottom line.[3]

The contractor just didn't have enough reserve to carry out the project while still complying with the many restrictions and difficulties that developed. So, in the final months, the performance bond that had been required as part of Terry & Tench's contract with the Bear Mountain Hudson River Bridge Company was called in. The Globe Insurance Company had underwritten the $1 million bond that would guarantee the continuity of the project, and there is confirmation that Globe was contacted sometime in June to issue a $1 million payment against the bond. George W. Perkins Jr., as cosigner, was ultimately responsible for the loan. Perkins soon entered negotiations with Globe regarding his obligation. In a letter to J. DuPratt White, Perkins explained, "My total outlay . . . amounted originally to $750,000. This is the round sum of money for which I settled with the insurance company. You will remember the bond was a million dollars. The amount which I was required to pay them was reduced because of Tench's collateral which they held and because of other considerations which entered into the situation."[4]

In addition, there was a negotiation between Perkins and Tench, resulting in Tench assigning several additional real-estate assets to Perkins to minimize the cosigner's losses. However, most of these assets were mortgaged, so Perkins had to first clear the mortgages, which cost him a total of $61,000. In addition, another piece of real estate appraised at $197,500 had a mortgage of $50,000, which Perkins again had to satisfy from his funds to liquefy the asset, and that amount, coupled with the $61,000, brought the total of Perkins's additional investment to $111,000 ($61,000 plus $50,000).

The assets appraised at $197,500, minus the $111,000 outlay, resulted in a profit of $86,500. Perkins then explained to White that after he deducted this profit from the $750,000 adjusted debt, he was left with $663,500 as an out-of-pocket number. Perkins goes on to tell White, "I have, however, Tench's further agreement to pay me half of his personal income in excess of $20,000 per year and the agreement of Terry & Tench to pay me half of the company's net income. Whether or not these agreements are worth anything, I am sure I don't know. So far, I have received no payments on their account."[5] As the cosigner, Perkins was ultimately responsible for the $663,500, but since Tench and his company were the primary debtors, Tench would reimburse Perkins directly, according to the agreed commitments, assuming the outstanding debt was paid in full.

Besides the financial difficulties, Tench had other issues to consider. On May 17, 1924, Tench's partner and good friend Edward Terry died from an unrelenting heart ailment, having suffered for several months before losing his battle. Terry left behind his wife, Rebecca Allie, and a son, Ezra David Terry. Tench was grief stricken. He and Terry had shared a partnership as well as a close friendship dating back several decades. The collaboration between the two men had begun in their early days of bridge building and culminated in the formation of their company, Terry & Tench Co., Inc., in 1895, and, of course, it continued as their success grew and their friendship developed. An obituary in the *Engineering News-Record* noted that many of Terry's colleagues considered him one of the most successful construction engineers in America at the time.[6] His passing left Tench with the sole responsibility of resolving the (many) outstanding issues and developing a plan to carry on. Finishing the bridge under the current circumstances would require enormous determination and fortitude. Tench would have to muster all his energy and determination to do so.

These developments, of course, affected the construction schedule. The job report dated July 1 alluded to some of the issues and reported that only minimal work was done on the bridge itself during the previous month. There were a small number of cable wires placed in June and some progress on the approaches. Excavation and grading on the east approach were now 90 percent complete, and work on the retaining walls in that area was under way. On the west approach, drilling and blasting were completed, allowing excavation to resume.[7]

Fortunately, things picked up again in July. That month McClintic Marshall shipped most of the remaining steel to the work site. By this time, Roebling had manufactured 100 percent of the cable wire and shipped approximately 88 percent to the site, with installation beginning sometime in June. The installation method would be perfected as the process moved forward. In planning for the fabrication of parallel wire bridge cables of large diameter by the aerial spinning process, several conditions must be considered. The size and total suspended length of the cables are important, but the number of strands, the number of wires, the type of wire connections, and the total tonnage of wire all play into calculating how long the spinning process will take. More specifically, the accessories to these items must be selected with consideration of all potential difficulties that may arise. In order to estimate the length of time it would take to complete the spinning operation, it was assumed that the first 25 percent of the wires strung would take twice as long as the second 25 percent. Once the first 25 percent is done, the men are usually comfortable with the process, and most will have reached their highest level of performance. The balance of the work should progress at the same rate as the second 25 percent of the work. The Bear Mountain Bridge is carried by 2 main cables, each cable is made up of 37 strands, and each strand contains 196 wires, for a total of 7,252 wires in each. Each of these cables measures 18 inches in diameter after compaction and wrapping. Each wire in these cables is 0.195 inches after galvanizing (application of a protective coating of chemically pure zinc) and has a measured strength of 215,000 pounds per square inch.

The spinning process for the Bear Mountain Bridge, simply explained, proceeded in the same way as that used in the Williamsburg and Manhattan Bridges. The reels of wire were placed at both ends of the bridge, and as individual wires were pulled out from one end of the bridge and anchored securely to eyebars within an anchorage, they were then taken by a traveling sheave or trolley and transported up and over each of the towers. At the top of a tower, the wires were supported on cast-steel saddles designed to seat the cables, allowing them to be secured by steel casting clamps. The towers were designed to allow for any longitudinal movement of the cables owing to temperature or live load stresses. The wire continued via the tramway wheel down to the opposite anchorage, where a workman received it and secured the end to an eyebar within that anchorage. A new wire would then be laid over the wheel, and it would start the return journey to the opposite anchorage. The sheaves were driven by 50-HP electric motors, one placed at each anchorage pit.

While the size of a bridge and the number of cables used in each impacted the time needed to install the main cables, as projects were completed and the process was streamlined, the results improved. Also, a bridge with longer spans allows more wire to be spun per day since the runs of the tram wheel would not be interrupted as often. On the Bear Mountain Bridge (1924), there were two main cables. The spinning process involved one tramway with two traveling sheaves. This resulted in four strands per cable being strung at one time. On the Williamsburg Bridge (1903), there were four main cables. The spinning involved one tramway with two traveling sheaves, resulting in only two strands per cable being strung at one time. On the Brooklyn Bridge (1883), there were also four main cables, but the wires crossed from one side only, which resulted in only one strand per cable being strung at a time. These factors play into why the rate of stringing was twice as fast on the Williamsburg Bridge compared to the Brooklyn Bridge and four times as fast on the Bear Mountain Bridge. The Brooklyn Bridge took twenty-one months to spin its cables, the Williamsburg Bridge took seven months, and the Bear Mountain Bridge, partly owing to the new flat-wire serving band, completed its cable spinning in just three and a quarter months, setting a record and accelerating a schedule that had maintained a constant delay factor almost from the start of construction.[8]

The best day of production for cable laying on the Bear Mountain Bridge was on July 28. There were 492 wires laid during a span of twelve hours, and, therefore, the projected date for clamping was moved up to

August 15. This development reduced the previously expected four-month delay by at least two and a half months. Additionally, updated projections regarding the excavation, coupled with the progress of grading on the entire eastern-approach road, set an expectation that they could be complete by October 1. In early August the concrete paving for the east approach was listed as 48 percent complete, while the paving of the west-approach highway was expected to be done by the end of that month.[9]

According to the September project report, McClintic Marshall had assembled all materials by September 2, except for one section of stiffening trusses that was expected the following week. The elevation at which the bridge deck would meet the east and west shores to maintain alignment with the roadway was determined to be higher than the minimum clearance dictated by the federal government. Therefore, the engineers decided to install the stiffening trusses beneath the deck rather than atop the deck, as was done on many other Hudson River crossings and East River bridges in New York City. The result was an unobstructed view of the majestic Hudson from either direction.[10] The stiffening trusses on the Williamsburg Bridge are forty feet high and were installed atop the deck along the perimeter of the bridge rather than below the deck; therefore, a pedestrian or motorist crossing the bridge is prevented from enjoying any view of the East River or the surrounding landscape. The decision to place the Bear Mountain Bridge trusses below the deck also allowed the main cables to connect to the lower member of the truss rather than the upper member, a difference that allowed for more stability in the deck installation.

By the end of August, Roebling had shipped and placed 100 percent of the cable wire, and 40 percent of the squeezing was done. The footbridge cables were not needed any longer and were released and installed as suspenders. The actual spinning of the wires, after the preparation work was completed, took only seventy-five days—from June 9 to August 24, 1924—accomplished by a crew of 159 men. As with any process, there was much that could have gone wrong, and even though the men took great care to minimize problems, delays and mishaps were a threat. The wire was stiff and subject to shifting off the traveling sheave, a problem that could require work to stop while the wire was replaced. Much of the work was done manually, and human error could not always be eliminated, no matter how careful workers were. Reels sometimes arrived at the job site incorrectly wound and would have to be rewound. Splices often came apart and had to be respliced. All these possibilities impacted the time necessary to complete the work.[11]

In September the concrete roadway at the west-approach span was completed, and the east-approach highway was noted to be 98 percent excavated and 60 percent paved by this time.[12] The job report dated October 1 notes that the following work had indeed been accomplished, as expected: All steel was on site, all cable squeezing was complete, the transfer of footbridges to cables was done, and the cable bands and suspender ropes were in place. Additionally, stiffening trusses consisting of nineteen panels on the east side and twenty-one panels on the west side—including bottom chords, web components, bracing, and floor beams of the suspended span—had been installed; taken together, these elements amounted to about 44 percent of the total steel in the suspended span. An extraordinary rate of speed was reached in the placement of the steel components at the main span. Components were shipped to arrive at railroad terminals located at the piers in New York City. From there, they were sorted and organized, loaded on barges and scows (a smaller type of barge used to carry cargo in the nineteenth and early twentieth centuries), and transported north to the site. Once they arrived, the men assembled the floor beams and the transverse framing members on the barge, then lifted the completed truss into position within the span.[13]

There was some talk about using precast concrete slabs if the steel erection was not complete before the cold weather arrived, since the weather would impact the pouring of concrete. The bridge company approved this change even though Baird was not in favor of using the precast. He deemed it unsuitable for a flexible structure such as a suspension span of narrow width. He eventually conceded in deference to the BMHR Bridge Company. Because of the efficiency of the steel erectors, the floor construction was completed and ready for concrete well within the time frame prescribed by the weather. Therefore, the precast idea was abandoned. By October 25 the concrete was poured, and the deck on the east approach was in place.[14] Both east and west anchorage tunnels, as well as the portal cuts in the east-approach rock face, were all finalized, and the east-approach highway excavation and related retaining walls were complete, with 32 percent of the parapet walls and boulder parapet in place. The momentum was encouraging, and as of early September 1924, those individuals involved with the construction started to believe it might be possible to open the bridge by the original target date of January 1, 1925.[15]

In marked contrast to the earlier months, so characterized by low morale, there was now a feeling of camaraderie throughout the site. The men felt a sense of pride in the accomplishment; it looked as though they might bring the project in on time after all. Despite the many obstacles encountered, they had prevailed! And not only did they bring the project together, but in the face of such extremely dangerous work there had been no job-site worker fatalities to mar the victory. However, within their tight-knit community, there was much sadness and regret over the loss of two of their fellow workmen who had been killed in nonworksite accidents while walking along the railroad tracks after their shifts had ended. Steve Bilovick, an employee of Terry & Tench, was found on April 14, 1924, on the tracks outside of Peekskill. It was determined that Bilovick was hit by a southbound train as he walked the tracks around eleven thirty at night. His remains were discovered by two coworkers who followed the same path from Manitou to Peekskill along the tracks about an hour later. The railroad investigation that followed confirmed that he was forty-six years old, had black hair and a mustache, and had a wife and children residing in Italy.[16] One week later, on April 20, 1924, another workman, Maries DeMastini, an employee of W. F. Carey and Co., Inc., was hit and killed by a northbound train as he walked on the tracks at the north end of the Annsville drawbridge early in the morning. DeMastini's remains were discovered by workers en route to the bridge construction at six thirty in the morning. Investigators reported that he was between forty-five and fifty years old and was from Readsboro, Vermont.[17] Both men were lodging in the area while they worked on the bridge. Both had families that, sadly, had to be notified of the tragedies.

On October 8, ceremonies were held at the site of the bridge to mark the first official visit of the Peekskill and Bear Mountain Bridge

Celebration Commission. Former New York governor Benjamin B. Odell, now vice chairman of the Celebration Commission, was invited to drive the final rivet into the bridge. Assisting him was John Herpt, a steelworker who was working on the bridge and had narrowly escaped death the previous day while dangling 153 feet above the river. Herpt, along with his coworker Joseph McKinnon, was riding on a steel girder as it was being lifted into place within the bridge framework. As the girder settled down, Herpt was caught against an adjoining upright steel pillar and was injured. The girder was only twelve inches wide, so as Herpt swayed, it seemed as though he might lose his balance and tumble into the river below, which from that height could have easily been a death sentence. But McKinnon, at the risk of his own life, reached out and put an arm around the injured man. For a few moments, people witnessing the incident thought that both men would fall into the water below. But McKinnon kept his balance and was able to hold on and steady Herpt against the upright. Shortly afterward, rescuers lowered them down to a launch. Herpt was unconscious as he was taken ashore. Both men were transported to the hospital for observation and treatment. Thankfully, they both escaped serious injury and were back on the job the following day.[18]

The job of a bridgeman is undoubtedly dangerous, and it was a triumph to have come through the previous twenty months without a serious incident. Newspaper accounts testify to accidents occurring at other job sites: some tell stories of survival from extreme heights, while others recount tragedy even from low elevations. A tale is told of a bridge worker who was considered a legend because in his career he had experienced more than thirty falls, all of which he lived to talk about—one was from a height of 87 feet from which he escaped with nothing more than a scratch. But another occurred while he was working on the elevated railroad loop being erected at the Brooklyn end of the Williamsburg Bridge, where his luck ran out. Although he fell only an estimated 40 feet, he hit the ground on his right side, and his head found a three-inch nail protruding upright. It entered his skull unnoticed since it was concealed by his generous mop of hair. While operating on him, the doctors were surprised to discover the nail embedded in his head. The newspaper article held out little hope for his survival.[19] Another bridge worker dropped 175 feet into the water below and swam to the shore with little to no consequence.[20] Ironworkers have earned the nickname "cowboys of the sky."[21] There appears to be no shortage of bravery among these men—personified not only by the midair daredevil performances they take in stride every day but also by the quick-witted, selfless instincts that were spotlighted in 1902, when the south tower of the Manhattan Bridge, still under construction, erupted into flames. The bridge crew was on site when the fire broke out. The firefighters arrived to fight the blaze, but they were unfamiliar with how to access the primary area of flames high above street level. Newspaper accounts document that the crew members jumped into action without hesitation, leading the firemen up the structure into the most dangerous areas of the fire.[22] "These bridgemen know the value of a man who takes his work and its danger most seriously, and who watches the man next to him—for in their business one man's error often means another man's life."[23]

By November 1, 1924, the BMHR Bridge Company reviewed the progress and once again made a change to the projections. Unbelievably, it was now looking like the bridge would be ready for traffic by November 26, 1924—a full month ahead of schedule! The suspended span was finished on October 15, accomplished in only thirty-four working days.[24] The concrete deck of the roadway was also complete, as were the east and west abutments and the west cable chambers. Sidewalks, railings, and asphalt work, which had been projected to take another three weeks, were completed by November 9. The wrapping of cables (which was not required to open the bridge to traffic) was deferred until the following spring. One coat of paint had been applied to the suspended span and railings, and it was decided that the remaining coats (also nonessential to the opening) could be done in the spring. It is a little-known fact that the original color of the bridge was black. In the early 1920s many of the metal paints were of darker shades, such as maroon, blue, green, or black. Carbon black pigment was one of the cheapest available, and the use of gilsonite, a form of bitumen, made the paint very durable. It was not until years later that the bridge's color was changed to gray or aluminum color.[25]

The eastern-approach highway concrete paving was set to be completed on November 1, and the west-approach highway was already finished up to the tollhouse. Smith was confident that the east tollhouse would be ready by November 26. Since the west tollhouse was not supposed to be completed until January 1925, plans were made to install a temporary structure before the opening;[26] however, it turned out that both tollhouses were completed and ready by November 26, 1924, after all, except for some minor finishing work that could be put off until after the opening. The creepers, derricks, and other equipment were removed by November 22. Although several small, inconsequential tasks remained, the site was cleaned and the bridge was cleared for traffic.

The remarkable progress made during the final two months, while facilitated by unusually favorable weather, was credited to excellent collaboration and correctional management. But it was also wholeheartedly agreed among all involved that the skillful execution and enthusiastic spirit exhibited by all the workmen were just as crucial to the acceleration of the schedule and the fulfillment of this amazing architectural accomplishment. Wilson Fitch Smith noted in his November 1 report that "such an accomplishment is very unusual on contract work."[27]

Plans for opening-day festivities had begun as soon as it was clear that the bridge would be ready by Thanksgiving Day. Invitations were sent out, and many well-known community leaders were asked to participate. On November 26, 1924, the bridge opened amid fanfare and speeches staged at the bridge and afterward at the Bear Mountain Inn. The West Point marching band provided the music, and reporters and photographers showed up to record the event for posterity. Roland and his staff tailored a program that would announce to the world the spectacle that was the Bear Mountain Bridge. Roland's mother, Mary, was asked to preside over the unveiling of the historic plaque labeling the bridge appropriately with its statistics and noting the date of the dedication. Honored guests were invited to speak, and a grand luncheon was planned to be held at the Bear Mountain Inn immediately following the ceremonies at the bridge. Roland acted as master of ceremonies for the festivities situated on the bridge itself, set up at the west-end tollhouse, and took the microphone as well at the luncheon that followed at the inn.

During the luncheon, Roland announced the names of those individuals who were instrumental in planning the bridge, overseeing the construction, and supervising the various crews that brought the bridge to life.

The following day the bridge opened to the public, and lines of traffic crossed the Hudson as local citizens sipped their morning coffee and read accounts of the previous day's festivities reported in local valley newspapers. Across the nation, the news was reported in major journals in large US cities (New York and others), as well as being reported in European publications—all singing the praises of the longest suspension bridge in the world!

The celebrations were over, and the festive music played by the West Point Band wafting over the deep, blue Hudson was but a memory in the minds of those persons who had attended the bridge's opening. All that remained were shreds of bunting that had decorated the bridge trusses and perhaps a few of the small international flags that may have escaped the cleanup committee's grasp. There were still miscellaneous tasks to be completed, even though the bridge was now in public use. Tench was still on the job, and he would complete these remaining items as part of his commitment. With the passing of Edward Terry, the company known as Terry & Tench Co., Inc., was no more. Tench dissolved the partnership and reorganized the company under the name of Tench Construction Co. Was this reorganization because of the company's financial problems or because of Terry's passing? It is unclear. Nonetheless, on January 7, 1925, Tench signed a new contract with BMHR Bridge Company under the name of Tench Construction Co. and continued to complete the miscellaneous items that needed to be done.[28] Although Frederick Tench worked for the remainder of his life, he never achieved the level of success that Terry & Tench had enjoyed in the years immediately preceding the building of the bridge. Frederick Tench died peacefully at his home in White Plains on October 27, 1944, one day before his eighty-second birthday. His legacy is characterized by his life's work in the building trade across the country, and his influence on the state of New York is highlighted in the completion of the Bear Mountain Bridge.[29]

Homage was paid to Frederick Tench in an article published in the *Yonkers (NY) Herald* following the bridge opening. Cornelius A. Pugsley, an investor and a director of the BMHR Bridge Company, gave credit to Tench for conceiving the bridge plan and for his persistence and courage in seeing the construction through to completion.[30]

On November 28, 1924, the *New York Times* published an article praising the Bear Mountain Bridge in poetic terms. The builders "invite[d] all America to come and be bewitched" by its concrete lanes and stunning setting, as well as its "approaches of surpassing beauty!"[31] It was a fitting finale to the blood, sweat, and tears of the previous twenty months, and it captured the sense of marvel the new bridge evoked. It was done!

20
The Bond

Still lingering, however, were the effects of the financial complications that developed for Tench and George W. Perkins Jr. The bond had ensured that the project moved ahead relatively unscathed, but, as cosigner, Perkins seemed to be the one left to deal with the residual effects.

Though the bond was called in 1924, the negotiation and resolution of the matter continued for several years. As Roland witnessed the evolving developments, he felt a responsibility and thought it was unfair for Perkins to shoulder all the liability. In 1928, he decided to consult with the investors he had personally solicited for the project and ask them to support their friend and colleague who, in support of the project, had agreed to cosign the note. In correspondence from Harriman to various bridge investors, he reasoned that since, at this early point, the investment was seeing a small return, it would not be a hardship for each of them to sign over a portion of their investment to Perkins, thus creating a "cushion" to soften his financial blow. Roland was dogged in his pursuit and execution of the effort. He wrote a letter to each associate within his circle, explaining the situation and strongly urging them to cooperate:

> Dear Mr. [recipient's last name],
> In 1923, at my personal request, you purchased . . . Income Bonds of Bear Mountain Hudson River Bridge Company . . . together with shares of stock. . . . I realize that it was primarily your sense of public spirit which induced you to make the purchase as I made it clear to you at the time that the bonds were highly speculative.
> The fact that during the past two years the Bridge Company has become a paying proposition resulting in the commencement of interest payments on the Income Bonds leads me to place before you

one phase of this matter, involving a real injustice, which I feel you will be willing to join with me in correcting, at least in part.

As one of the conditions leading up to the letting of the contract for the construction of the bridge, the contractor was required to furnish a performance bond guaranteeing completion of the bridge within the contract price. The surety companies that furnished the bond demanded from the contractor an additional name on the back of the bond and, accordingly, in a generous, public endeavor to see the bridge project successful, Mr. George W. Perkins agreed to stand behind the contractor and signed an indemnity agreement to the surety companies. For various reasons, the contractor failed to perform satisfactorily and the Bridge Company was forced to call upon the surety companies to complete the bridge. After extended negotiations, a settlement was finally reached by which the surety companies paid to the bridge company a flat sum of one million dollars ($1,000,000). At the same time, Mr. Perkins made a settlement with the surety companies which, I understand, involves a net loss to him of something over six hundred and fifty thousand dollars ($650,000). Now that the others of us who were motivated by the same spirit as Mr. Perkins, I believe have our investment secure and can look forward to a very probable profit, it occurred to me that you and a number of the other Income Bondholders would be entirely agreeable and would welcome the opportunity to join with me in contributing a portion at least of our prospective profits toward the partial recoupment of Mr. Perkins' losses. Therefore, I make the proposal that you join with me in assigning to Mr. Perkins two shares out of each five shares originally received by you.

I have discussed this matter personally with two or three of the larger Income Bondholders and they are sympathetic with the plan provided that a sufficient number of other Income Bondholders join it. If all those to whom I personally sold Income Bonds join in this plan to the extent outlined the resulting number of shares contributed would be about 2,000, or about 18% of the total capital stock. It is quite clear that this will by no means recoup Mr. Perkins for his losses but it will give him the possibility of making up part of the losses and in fairness it seems to me the least that we can do. You will also appreciate that if the State should take over the Bridge at the end of the first five-year period, there will be practically nothing for the stockholders after taking care of all income bond interest and providing for the retirement of First Mortgage and Income Bonds at a premium of 5%.

I will appreciate your early response as I am leaving New York toward the end of this month and would like to settle this matter before my departure. In any event, I am prepared to assign to Mr. Perkins my proportion under the foregoing plan and I hope that you will join with me, although, of course, there is no obligation on your part to do so. If you have any suggestions as to any other form this adjustment might take I will be delighted to have you express them.

Sincerely yours,
E. Roland Harriman[1]

The responses recorded seem to indicate that all reacted very positively to the request and expressed that they were happy to contribute to Perkins's unfortunate predicament, but not everyone penned their reply promptly. Accordingly, Harriman repeatedly found himself writing follow-up notes. Some correspondents were traveling abroad, whether on business or pleasure; several were ill and indisposed; and some were dealing with business-related issues—all seemingly legitimate reasons for not responding quickly.[2] Nonetheless, he continued unabashedly to write to each of them, until he had received from each man the requested stock certificates, which he would then transfer to Perkins. This dedication is a testament to Roland's loyalty to his friends and associates. His persistence in continuing the follow-up letters and tracking down the investors until an answer was delivered from each demonstrates his determination to do what he felt was right.[3]

The back-and-forth between Roland and the other investors commenced in August 1928 and was not fully resolved until sometime in 1930. In all, nineteen of the original investors agreed to come to the aid of the cosigner. Besides Roland Harriman, some of those who agreed to sign on included John D. Rockefeller Jr., L. V. Ladoux, G. W. McGarrah, H. L.

Satterlee, Mortimer L. Schiff, and Paul and Felix Warburg, to name a few well-known investors. With each man agreeing to sign over 40 percent of their stock holdings, Roland was eventually able to present Perkins with 1,422 shares of stock to reduce his exposure.[4]

Harriman's plan involving the investors' contributions was delivered to Perkins in August 1929. Perkins was delightfully surprised and responded in a letter to Roland, expressing his amazement and full appreciation—not only for Roland's contribution of shares of stock, but also for the efforts he expended in approaching the others, organizing the plan, and administering it to completion. "I know that you feel that fundamentally we all went into the Bridge from the same point of view and the loss which I incurred ought to be spread as much as possible, which is mighty fine, but it doesn't change my feeling at all that what you have done is perfectly great," he wrote to Roland.[5]

Along with Roland's letter, Perkins received an accounting of all investors involved, listing their names and how many shares each had contributed. In his response, he indicated that he intended to write to each one of them to express his appreciation.[6] The dividends paid on these 1,422 shares did not fully compensate Perkins for his exposure, but they did help to minimize his loss. If the bridge continued to be successful in collecting toll revenue, the payments would continue over several years.

The Bear Mountain Bridge was now a permanent feature of the Hudson Valley landscape. Residents boasted that it was the longest suspension bridge in the world[7]—even if that distinction lasted for only nineteen months, until the Ben Franklin Bridge, which crossed the Delaware River from Philadelphia to Camden, was completed and assumed the title. As each new monument was built slightly longer, each claimed to be the "longest span in the world," until a new challenger appeared and usurped the title for itself.

"The Golden Age of Bridges" is a term used to label the forward-looking infrastructure movement that swept the country during this time. However, exactly when it began and when it ended is a matter of contention. In 1883, the building of the Brooklyn Bridge seemed to have initiated the movement in New York, introducing several new methods of bridge building. While there were several smaller bridges built in the years following, there was no bridge of any consequential span built for the next twenty years. The Williamsburg Bridge appeared in 1903, and the Manhattan Bridge followed in 1909 along with the Queensboro. The Bear Mountain Bridge, built in 1924, utilized and reintroduced many of the experimental practices used in the Brooklyn Bridge, and there is a school of thought that considers the Bear Mountain Bridge to be the spark that reignited New York's Golden Age of Bridges, which would continue until the beginning of World War II.[8] Within the decades following the Bear Mountain Bridge, a parade of new civic monuments appeared, fostering a steady development of new engineering practices. The George Washington Bridge was built in 1931, seven years after the Bear Mountain, and the Triborough Bridge came five years later in 1936; the Bronx-Whitestone was built in 1939.

Soon after, World War II monopolized the attention of the country, and many plans for infrastructure projects were interrupted. In 1955, during the Korean War, the Tappan Zee Bridge was built as a cantilever structure joining Rockland County and Westchester County. Because of material shortages occurring at the time, the bridge was built with the expectation of lasting only fifty to fifty-five years. Therefore, in 2019, a new cable-stay bridge was constructed to replace the 1955 structure. Several years after the first Tappan Zee was built, the Throgs Neck Bridge was built in 1961, and the Verrazzano-Narrows Bridge followed in 1964 with a span of 4,260 feet, which was the longest span in the world at the time, and still to this date this bridge holds the record as the longest suspension bridge in North America.

During the early years of operation, the Bear Mountain Bridge provided a small return to its investors, but the finances soon began to decline.[9] Within a few years of its completion, the original investors watched as the revenue from the bridge sputtered, stalled, and receded to a point where it was barely supporting itself and its list of necessary maintenance operations.

One cause of the decline in revenue was the competition. Within the decade following the bridge's opening, the George Washington Bridge opened (1931), a crossing much closer to New York City, offering a better option for travelers seeking a route to Manhattan and providing a shortcut in distance, time, and costs.

Access to the parks on the western shore continued to generate traffic, but reports of outbreaks of typhoid fever across the country began to appear in the newspapers, and when a local flare-up was reported in a camp located inside Bear Mountain Park,[10] attendance levels dropped and, with them, the amount of bridge traffic. Other seemingly unrelated factors also affected the bridge's traffic numbers from as early as the months immediately following the bridge's opening. Not long after the bridge opened, a road project on the eastern side of the river was undertaken to upgrade the Albany Post Road. The construction and resulting delays discouraged automobile travelers from taking the route that would have ultimately brought them to the bridge, since they would first have to navigate roadwork delays. The cumulative effect was an overall decrease in traffic across the bridge, and therefore a decrease in revenue.

To fight the steady decline, the Bear Mountain Hudson River Bridge Company tried several different ideas early on. One of these involved marketing the bridge as a better route to the marketplace. They reached out to those individuals in the business of selling and transporting produce from area farms to the city, pointing out that the forty-five-mile route from the bridge to Manhattan provided the clearest access, with no impediments or obstructions, especially if the travel was done at night.[11] The investors also sought out methods of reducing their normal operating costs and were successful in refinancing the debenture bonds at a lower interest rate, all of which helped, but did not have a significant impact on the overall revenue generated.[12]

By the end of the "Roaring Twenties," the country was immersed in the Great Depression, and the Bear Mountain Hudson River Bridge Company began to acknowledge that their investment in the bridge would probably never provide the return they had originally projected. However, they continued to monitor the operation closely, keeping expenses minimal and revenue as high as possible.

The Palisades Interstate Park Commission continued as manager of the parks, all the while developing many innovative programs. The Bear Mountain Inn became a popular meeting place for cultural celebrities as well as average residents.

21

Remembering

In June 1928, the New Jersey State Federation of Women's Clubs once again reached out to the PIPC on the status of their memorial to be erected within the Palisades Interstate Park. A letter regarding the matter, written by Lydia S. Osborne, the chairman of the federation's Memorial Committee, found its way to the desk of William A. Welch, the general manager of the PIPC. Welch responded to Osborne's concern over the continuing delay, and with that correspondence, he included three potential designs for a monument that he had sketched. He asked for some feedback from the women's committee as to their preferences. Each of the designs, done by Welch's hand, included a small rectangular base with a tower rising upward.[1] Osborne's reaction to the general manager was very positive, indicating that the ladies were delighted. They selected the design that included a round tower and asked if a staircase could be incorporated to allow the visitor access to the top level where they might enjoy the view. In July Welch responded, saying he was glad that the board was pleased with the plans. He promised to provide more details as soon as possible. On September 24, Welch wrote to Osborne again, explaining that to contain costs the construction of the monument would be done by "employees of the Park" who were "currently immersed in erecting a new administration building for the park in Alpine, New Jersey." He added that there were delays in the construction of the administration building owing to weather conditions and that the crew would not be ready to start the monument until the current project was completed. Therefore, he could not commit to a date at that time. Given the uncertainty of a schedule, he cautioned the women not to expect the memorial to be done before the following spring. He closed with a question on the preference of the committee regarding the roof style for the tower.[2]

The following spring, on April 30, 1929, the long-awaited Watch Tower was dedicated within a three-acre memorial park honoring the New Jersey State Federation of Women's Clubs and those persons who had labored to effect the creation of the Palisade Interstate Park Commission in 1900. A printed event program outlining the festivities of the day noted that Frederick C. Sutro, president of the PIPC New Jersey Commission, presided over the ceremonies along with J. DuPratt White, president of the PIPC New York Commission, and spoke to several hundred attendees. Present on behalf of the NJSFWC were Mrs. Louis V. Hubbard (then president of the federation), Mrs. S. Elizabeth Demarest, Mrs. Cecilia Gaines Holland, and a large contingent of federation members. The program also noted the presence of several PIPC staff members in addition to Sutro and White.

Cecilia Gaines Holland, the only one of the original crusaders who was listed as attending the ceremony, delivered the main address of the day. Her speech emphasized the history of the movement and proclaimed that at the turn of the twentieth century, the women being honored were "then not even citizens. We had no vote, no power, either political or financial. What we did have was enthusiasm for the shores of the most picturesque river in the world, and that love of conservation which has always been a characteristic of women. It is to mark these first efforts that this memorial is raised." She went on to note that the memorial was "not the real monument. The real monument is the Palisades themselves, standing in majestic grandeur, saved from destruction by patriotic men and women of New York and New Jersey."[3]

Vermilye, Sterling, Loomis, and Holland were still living (Sauzade had died in 1918, at the age of eighty-six), but Holland was the only name listed in the program as having attended the ceremony. Whether Vermilye, Sterling, or Loomis was present is not known, nor do we know what their state of health and mobility was at the time. A "casket of records" containing memorabilia of the fight to save the Palisades was placed in the cornerstone within the Watch Tower, and a bronze plaque was affixed to the face of the tower with the following inscription: "This Federation

Memorial Park is dedicated to the successful efforts of the New Jersey State Federation of Women's Clubs and of those men and women who aided in the opening years of the twentieth century in preserving these Palisades Cliffs from destruction for the glory of God who created them and the ennobling of the generations which may henceforth enjoy them."[4]

It had taken almost thirty years, but an acknowledgment of appreciation for these noble women had finally been made.

Adaline Wheelock Sterling died the following year, on November 25, 1930, at the age of eighty-three, and is buried in Greenwood Cemetery in Brooklyn.

Three weeks later, on December 18, 1930, Elizabeth Breeze Vermilye died and was laid to rest beside her parents in the family plot at the Brookside Cemetery in Englewood.

Sarah Sophia Dana Loomis served as an enthusiastic member of the Women's Club of Englewood and of the NJSFWC until her death. She died on September 5, 1939, in Englewood and is buried in Mount Pleasant Cemetery in St. Johnsbury, Vermont.

In March 1943, Cecilia Holland relocated to Oakland, California, after the death of her husband, John, to be closer to her daughter. Her California residency was short-lived; Cecilia died on June 30, 1943.[5]

By 1934, William Welch and the PIPC staff completed another memorial to a major benefactor for the Palisades Interstate Park Commission. This time it was to honor George W. Perkins Sr. On the day of its dedication, the president of the United States (and former governor of New York), Franklin Delano Roosevelt, accompanied by Eleanor Roosevelt, traveled to the Palisades to dedicate the George W. Perkins Memorial Highway that leads to a sixty-foot granite tower high on the summit of Bear Mountain, overseeing the entire park. Both the highway and the tower stand as memorials to George W. Perkins Sr.—tributes to his endless efforts to conserve the area for future generations. As the years have passed, the memorial highway and the tower it leads to have come to represent not only the contributions of the senior Perkins but those of his son George W. Perkins Jr. as well. The memorial continues to attract thousands of park patrons each year, personifying an appreciation of these two dedicated men. The dedication of this father and son, and the entire Perkins family, to the Palisades Interstate Park Commission as well as the communities it serves has solidly connected the Perkins name to the Hudson Valley, and it is difficult to discuss one without including the other.[6]

During my research, I stumbled upon a multipage dissertation on the engineering achievements of the Bear Mountain Bridge, its history, and various details of its construction, written shortly after the completion of the bridge. On the fourth page of the paper, the writer identifies himself as the one who was retained to design the bridge and supervise its construction—none other than Howard Carter Baird. The document contains a treasure trove of information about the technical details of the bridge from one who was intimately acquainted with them.

In the document, Baird mentions that the Bear Mountain Bridge site had been proposed numerous times as a location for a bridge, and, in fact, sometime between 1912 and 1914, Baird and his mentor, Henry W. Hodge, had themselves been retained to prepare a design for yet another bridge at the site, a steel-arch bridge supporting a double-track railway. The project followed the fate of all the others, as it never moved beyond the preliminary planning.[7]

Baird's dissertation states that the Bear Mountain Bridge was a wire-cable suspension bridge with a span of 1,632 feet, and, at the time of its completion, it had the distinction of being the longest suspension span in the world. He also explains that the Bear Mountain Bridge had several unique qualities: "Besides being the longest suspension span in existence, [it was] probably the first long suspension bridge with concrete floor slabs, and was the first long suspension bridge for which the natural rock present at the site was utilized for anchoring the cables, in place of building masonry anchorages for that purpose . . . [and] the elevations of the approach roads and the navigation clearance fixed by the War Department made possible a deck structure which gives an uninterrupted view from the roadway, and what is more important, has torsional rigidity due to complete transverse bracing."[8]

Baird goes on to mention the critical attacks on the bridge's proposed design that appeared in local papers before construction: Lawrence Abbott referred to the bridge as a piece of "tin frumpery." But Baird refutes this assessment with his confident position: "As the Manhattan Bridge and the Williamsburg Bridge, and the Philadelphia Bridge [then in the planning stage] were designed to have steel towers, it seemed entirely proper to adopt such construction, provided proper proportions be used and simplicity and unobtrusiveness be the keynote. The absence of ornamentation and the omission of finials and bulky housing over the saddles contributes to comeliness, and the towers, since completion, have not called forth the criticism that might be expected had the above attacks been justified."[9]

He did concede that there was one incongruous detail, however, but it was not of his making. As a result of the PIPC's requirement for a "tunnel-like" element to be added to the original design (an effort to improve the look in response to those early criticisms of the design), a curved piece of steel was inserted into each of the towers to create an arch. This addition was, in Baird's opinion, incompatible with the original design, but he complied with the request.[10]

The document goes beyond just discussing aspects of the design, taking time to describe the loads, materials, foundations, and towers, and providing technical engineering and construction details, for readers interested in them. The unique "natural" anchorages are then discussed in great detail, since, as Baird mentions, the natural rock of Anthony's Nose on the east and Bear Mountain on the west was utilized to support a compression strength of more than five million pounds. Further topics included the temporary footbridges and the main cables, both provided and installed by John A. Roebling's Sons Company. The two were interconnected: the cables that were originally installed to be used for the footbridges (necessary during construction) were eventually dropped once the footbridges were no longer needed, and those cables then became the suspender cables that connected to the steel framework and stabilized the main cables. The final topic was the main span and details of the design and the load requirements.[11]

During the twenty months of construction on the Bear Mountain Bridge, Baird worked side by side with Wilson Fitch Smith, the chief engineer for the Bear Mountain Hudson River Bridge Company, as well as with consulting engineer William H. Burr, who was considered at the time to be one of the greatest engineers of the post–Civil War era. Burr was a professor of engineering at Rensselaer Polytechnic Institute, as well as at Harvard and Columbia Universities. Internationally reputed, he had worked on such projects as the Holland Tunnel, the Croton Aqueduct, the Panama Canal, and other famous infrastructure projects. Burr and Baird knew each other from the Phoenix Bridge Company as well as several projects that Burr had consulted on with Boller and Hodge. Wilson Fitch Smith had been a student of Burr and was a graduate of Columbia University's School of Mines. He also supervised the construction of the Kensico Dam and related structures.

William A. Welch, the general manager and chief engineer of the Palisades Interstate Park Commission, joined the bridge project, representing the interests of the PIPC and the state of New York. Other engineers involved included John M. Belknap, Victor Leopold, L. L. Morton, and Holden D. Robinson, who had served in 1903 as assistant engineer in the construction of the Williamsburg Bridge across the East River. An engineer with the John A. Roebling's Sons Company at the time of the Bear Mountain Bridge, Robinson was an expert on cabling. Together with Washington A. Roebling and Charles Sunderland (the company's chief engineer), Robinson had developed and tested new cable-spinning theories and methods. During the construction of the Bear Mountain Bridge, he perfected a wrapping for cable strands and machinery for binding the cables with wire that proved extremely successful and allowed the cabling work to be accelerated and the bridge to be completed ahead of expectations.

Baird was present at the opening-day ceremonies of the Bear Mountain Bridge on November 26, 1924, riding along with Tench, as noted earlier in our story. The festivities ended early for the men when the automobile they were traveling in broke down on the bridge they had labored over. They missed the lavish luncheon altogether, and, therefore, E. Roland Harriman expressed his disappointment in a letter to Baird, dated November 29, 1924:

My dear Mr. Baird:

I very much regret that you were not at the luncheon after the opening of the Bear Mountain Bridge on Wednesday, because I wanted to publicly express to you the appreciation of the Bridge Company and my own sincere personal congratulations upon the successful completion of the Bridge. I made distinct reference to it at that time but I am afraid it lost a great deal of force by your absence.

We realize the successful construction depended upon accuracy of detail as well as correctness of the general design, and we feel that due to your close attention to the details the structure is well built and suited to the duty it will be called upon to meet.

It may interest you to know that everyone I spoke to on Wednesday commented on its beauty and stateliness.[12]

Baird responded several days later, thanking Harriman for his kind words and stating, "An engineer, to whom a structure such as this is entrusted . . . finds his reward in his own consciousness of a duty performed, and in such recognition as may be voiced by others."[13]

By the late 1920s, Baird enjoyed the deep respect of his associates and was considered an expert in his field. Additionally, his many successful projects, so visible to a public awestruck by the new, previously unimaginable displays of technological achievement, were a credit to the discipline of engineering, which had only somewhat recently become professionalized in the United States. The first attempt to organize civil engineers as a professional society occurred as early as 1828 in Ohio when a group of engineers followed the lead of the Institution of Civil Engineers in Great Britain. Attempts to follow the English lead were made among engineers across the country, but none was successful until 1877.

However, in 1848, the Boston Society of Civil Engineers was formed, and, in October 1852, an effort was made to organize a society of civil engineers and architects in New York. But it was not until 1877 that the members finally took the formal legal steps to charter and incorporate an organization in the state of New York under the name of the American Society of Civil Engineers (ASCE).

Baird was a member of several engineering organizations: the Engineers Club of Philadelphia, the ASCE in 1898 as an associate member, and then as a corporate member in 1904. In 1917 he joined the American Institute of Consulting Engineers, serving as a member of the organization's council for three years and vice president for one.

In 1929 Baird had traveled to the United Kingdom to submit his application for admission into the United Kingdom's Institution of Civil Engineers. The institution was highly esteemed, and affiliation with it enhanced one's résumé as an accredited engineer. The documents prepared and submitted by Baird, as part of his application, offer us in-depth details about his early career experience at the Phoenix Bridge Company, as well as at Boller & Hodge. From these documents, dated January 17, 1929, and written in Baird's handwriting, we get an illuminating view of his journey from draftsman to engineer. Baird notes that in 1892 he served as a "draughtsman" on railway and highway bridges; by 1895, he points out he had been promoted to assistant engineer, designing those same types of bridges. By 1901, he was preparing shop drawings and design details on projects involving twenty-five thousand tons of bridge work with spans of up to 370 feet.[14]

Howard Carter Baird never married, and, as many bachelors do, he mainly kept to himself, living a private life with his career as his main focus. Apart from his membership in several engineering associations and institutes, there is little evidence of other social interaction, except for his membership in the Century Club. A private club in New York City and an important social hub for Baird, the Century Club was founded in 1847 by poet William Cullen Bryant and a group of his associates to support an interest in fine arts and literature. It originally focused its membership on artists, literary men, scientists, physicians, officers of the US Army and Navy, members of the bench and bar, engineers, clergymen, representatives of the press, merchants, and "men of leisure."[15] Some notable members over the years have included painters Asher Durand and Winslow Homer and architects Stanford White, Calvert Vaux, and Frederick Law Olmsted.[16] The club, which is still active today, maintains an art collection that includes works by many Hudson River School painters, some of whom were members. Being established strictly as a gentlemen's club, women were not accepted as members until 1989, and then only after an intense legal battle.[17] Today the Century Association (as the club is now known) is located at 7 West Forty-Third Street in Manhattan.

Baird was elected as a Centurion on November 18, 1927, and the event was recorded in the Century Club's candidates' book, as was the custom. The entry documenting Baird's election lists Edwin S. Jarrett as his primary sponsor and Edward P. Casey as his second. Both were engineers and members in good standing and signed the book presenting their colleague for membership.[18]

As Baird advanced in years, "the Century" (as members referred to it) became more and more important in his life. It was a keystone to his existence. He spent a great deal of time there playing bridge and simply passing time with friends and cohorts. His life had a certain order, and he tended to do the same things in the same way each day.

In 1956 Howard Carter Baird left his beloved New York City and relocated to Evanston, Illinois, to be near his nephew James Baird Jacob, the son of his sister Martha, so that he could receive proper care in his twilight years.[19]

Baird died on December 5, 1957, in Highland Park, Illinois, just one month short of turning ninety years old. He is buried in the Baird family plot in Cave Hill Cemetery in Louisville, Kentucky.[20]

29. Bear Mountain Bridge, circa 1930s. Courtesy of the Palisades Interstate Park Commission Archives.

30. Women's Memorial Tower at the dedication ceremony, April 29, 1930. Courtesy of the Palisades Interstate Park Commission Archives.

The Commissioners of the Palisades Interstate Park and The New Jersey State Federation of Womens Clubs

dedicate a

MEMORIAL WATCHTOWER

Federation Memorial Park
Alpine, New Jersey
Tuesday, April 30th, 1929

31. Program cover for the Women's Memorial dedication. Courtesy of the New Jersey State Federation of Women's Clubs Archives.

New Jersey State Federation of Women's Clubs

"This Federation Memorial Park and this Memorial structure are dedicated to the successful efforts of The New Jersey State Federation of Women's Clubs and of those men and women who aided in the opening years of the twentieth century in preserving these Palisades Cliffs from destruction for the glory of God who created them and the ennobling of the generations which may henceforth enjoy them."

History

1897. An all day meeting was held in Englewood by the Federation to study the problem of saving the Palisades. Mrs. Katherine Stanside figured on the program. Mrs. John Gifford was later made chairman of a committee for that purpose.

1898. Mrs. Holland and Mrs. Williamson interviewed Governor Voorhees. After much agitation the Governor appointed a commission to study possibilities for "Saving the Palisades." Mr. Franklin Hopkins was president and Mrs. Holland and Miss Elizabeth Vermilye were members of the original commission.

1900. League to Preserve the Palisades formed under leadership of Miss Elizabeth Vermilye, chairman of Forestry of N. J. S. F. W. C.

1908. Memorial Park dedicated to the New Jersey State Federation of Women's Clubs by Palisades Interstate Park Commissioners and money for memorial entrusted to the Commissioners by Mrs. Henry H. Dawson, President of the Federation. The money was raised equally by the Federation and The League to Preserve the Palisades.

Program

PRELUDE . . . . . . Alda Quartet

Mr. Frederick C. Sutro, President of the New Jersey Commissioners of the Palisades Interstate Park, presiding

THE STAR SPANGLED BANNER . Alda Quartet

AMERICA, THE BEAUTIFUL . Assembly Singing

THE VISION . . Mrs. John A. Holland
President of New Jersey State Federation of Women's Clubs in 1897 and member of original Commission

FULFILLMENT . . Mr. J. De Pratt White
President of the New York Commissioners of the Palisades Interstate Park

ADDRESS . . Honorable Morgan F. Larson
Governor of New Jersey

RECESSIONAL . . . . Alda Quartet

PRESENTATION OF MEMORIAL AND
UNVEILING OF TABLET Mr. Frederick C. Sutro
Representing the Commissioners

ACCEPTANCE . . . Mrs. Louis V. Hubbard
President of the New Jersey State Federation of Women's Clubs

AMERICA . . . . Assembly Singing

POSTLUDE . . . . . Alda Quartet

32. Program pages for the Women's Memorial dedication. Courtesy of the New Jersey State Federation of Women's Clubs Archives.

22

Update—1927 through the Present

Over the next several years the PIPC experienced many changes, as the original board of commissioners, one by one, began to step down. The old guard gave way to the new. New commissioners stepped up, ready to carry on the work. In 1938 J. DuPratt White resigned after thirty-eight years of service, leaving the presidency because of his declining health. He died in July of that same year, just two months after his resignation. Edmund W. Wakelee succeeded White as president and served through the war years until he died in 1945. William A. Welch had served the commission for twenty-eight years as chief engineer and general manager by 1940 when he stepped down from his formal position, but he continued to serve in an auxiliary capacity until he died in 1941. Frederick C. Sutro served as a commissioner from 1913 to 1931, then transitioned to executive director, serving in that capacity from 1931 to 1940, during which term he successfully steered the PIPC through the Depression years. Sutro left the PIPC in 1940 but went on to serve as president of the New Jersey Parks and Recreation Association for another twenty years, retiring at age eighty-five. After the retirements of Welch and Sutro were announced, Kenneth Morgan, a protégé of Robert Moses and a former superintendent of Jones Beach State Park, was installed as chief engineer and general manager of the PIPC.[1]

In 1942, soon after the United States had entered World War II, George W. Perkins Jr. left his position at Merck to accept a post with the Chemical Warfare Service of the US Army. Years earlier, at the end of World War I, he had exited the military holding the rank of second lieutenant; he reentered the service during World War II as lieutenant colonel and was promoted to the rank of full colonel shortly thereafter. During the war, he served in both European and Pacific theaters of combat, as well as in Washington, DC, and he was awarded the Legion of Merit medal for his service.[2] After World War II ended, George returned to Merck and Co. for a short time until he was once again summoned by the US government. This time he was called to Paris to head up the industry division of the Economic Cooperation Administration, an agency established in 1948 as part of the Marshall Plan.[3] The ECA was formed to ensure the rehabilitation and development of heavy industry in European countries affected by the war. Specifically, the ECA helped the countries under its jurisdiction manage exports, ensuring a balance of products in the marketplace, so that both shortages and surpluses were avoided.

In May 1949 President Truman reached across party lines and appointed George as assistant secretary of state in charge of European affairs even though George had been a prominent Republican for many years and Truman was a Democrat.[4]

As a commissioner of the Palisades Interstate Park Commission, George invested much of his time and energy serving Hudson Valley while attempting to balance those responsibilities with his governmental appointments and his work at Merck and Co. He soon realized that he must reassess his time and dedication and make some changes. As his father had done before him, he decided to step away from his private-sector responsibilities to dedicate himself fully to the service of his community and his country. In June 1948, he resigned from Merck and Co., and in August 1949, in his capacity as assistant secretary of state, he was selected to be the sole representative of the United States to an international group seeking to form a defense coalition of twelve nations; today this coalition is known as NATO (North Atlantic Treaty Organization). Several years later, in 1955, President Eisenhower made him the permanent representative of the United States to the NATO Council, with the rank of ambassador.[5]

In 1945, following in his father's footsteps, George Jr. was elected by his fellow commissioners to the office of president of the PIPC. The war was over, and it was assumed among the commissioners that the Palisades Interstate Parkway project, which had been halted after the United States entered the war in 1941, would once again become a priority. There was much enthusiasm and many positive expectations among the commissioners, many of whom were recently appointed. In November 1935, John D. Rockefeller Jr. had gifted to the commission acreage along the crest of the Palisades, with the specific intent that the property would be used as part of the parkway project, a welcome addition to their stewardship. Because of the expectations of the Rockefellers in connection with the gift, it was imperative in the minds of the commissioners that the project be reactivated as soon as possible. However, as the project was resurrected, so too was the controversy instigated years earlier by the Lamont family, a landowner near the parkway's proposed route. The years following were filled with many obstacles and disputes. In the end, however, George Jr. was successful in guiding the commission to victory. He dissolved the Lamont-Burnett opposition and calmed the fears of the adjacent townships. The final five-mile phase of the Palisades Interstate Parkway, which composed forty-two miles in its entirety, opened to the public in August 1958.[6]

The Perkins family exerted a great influence on the valley and the world. George Sr.'s respect for nature coupled with his love for his fellow man allowed him to structure the PIPC as an agency that developed the valley while at the same time protected it. George Jr. grew up in the shadow of his father, following his example of community service, and ultimately became not only a leader within the PIPC and the Hudson Valley, but, through his work in the State Department, a leader of his country and impacted the entire nation and, through the creation of NATO, the entire world.

A strong belief in the value of service ran in the Perkins family. Years earlier, his father had explained the family's philosophy: "I have long felt that it is not wise to leave all our public affairs to the politicians and that businessmen of sufficient leisure and means should for patriotic reasons give their attention to great public problems."[7] Judging from his lifework, George Jr. felt the same.

On January 10, 1960, George W. Perkins Jr. suffered a heart attack while at home with his family and died.[8] In an address he had delivered at Hill School in 1946, he implored the young men in his audience to make their mark on the world. "There has never been a time when there was a greater need in the world for intelligent, honest, and self-sacrificing leadership based on Christian principles," he said, describing qualities that, in the eyes of many, he himself personified.[9]

The Harriman family also left its mark on the Hudson Valley. While George W. Perkins Sr. was immersed in the fight to save the Palisades, E. H. (Henry) Harriman entered the historical landscape through a different portal. Harriman's initial land purchase in 1886 and his subsequent efforts to develop and maintain the property created a momentous influence on the area and the quality of life for those individuals populating it.

His contributions continued even after his death, when in 1910 his wife, Mary, donated land to the state of New York, following the wishes of her late husband. Admittedly, this philanthropy was more likely the result of his anger and dissatisfaction with the state of New York's plan to locate a prison within a stone's throw of his family estate than it was a gesture of civic duty. Edward Henry Harriman was a powerful man used to having things done his way. His manipulation resulted in not only a successful execution of his wishes but also the laying of a foundation for a legacy that has lasted for more than a century[10] and the impetus for a system of parks renowned across the nation.

Mary, to whom Henry left his entire fortune, proved to be a savvy manager of both the family's wealth and its legacy. Philanthropically minded, she reviewed and studied each charitable request, of which there were many once the public realized the size of the family's wealth. She carefully selected only those requests that she determined were worthy of being affiliated with the Harriman name. A strong sense of noblesse oblige was instilled in the Harriman children at a young age, and, as they grew to adulthood, each, in turn, assumed their place as responsible members of the community, serving in various capacities in philanthropic and socially minded organizations. Mary passed away on November 7, 1932. She was buried beside her beloved Henry in St. John's Cemetery at Arden.[11]

Mary Harriman Rumsey, eldest daughter of Henry and Mary, founded the Association of Junior Leagues in the early 1900s, and in 1933, she was appointed chair of the Consumer Advisory Board of the National Recovery Administration by President Franklin D. Roosevelt. At the same time, Mary worked with Frances Perkins, secretary of labor and the first woman to occupy a cabinet post (under FDR), to advocate for the enactment of the Social Security Act. It was owing to her encouragement as well as her social contacts that her brother Averell was drawn into the political scene, and during FDR's New Deal, he secured his first government job and went on to hold a variety of government positions. Mary died in 1934 as the result of an equestrian accident while on a hunt in Virginia.[12]

William Averell Harriman was a businessman, banker, shipbuilder, and politician. Averell was probably the most colorful of the Harriman children. He married three times, traveled around the world, moved in high-society circles, and enjoyed close relationships with many powerful political leaders. His first initiation into politics was in 1934 with his appointment to the National Recovery Administration, the centerpiece of FDR's New Deal, and subsequently to the Business Advisory Council. During World War II he entered foreign service when he was appointed ambassador to the Soviet Union by President Roosevelt. Under President Truman he was ambassador to the United Kingdom and afterward secretary of commerce from 1946 to 1948. In 1955 Averell was elected governor of New York and served until 1958 when he was defeated by Nelson Rockefeller. Under President John F. Kennedy he served as assistant secretary of state for East Asian and Pacific affairs and then undersecretary of state for political affairs. He remained active politically for several years and was recognized as an accomplished foreign policy negotiator. He entered the presidential races in 1952 and 1956, but both times he was unsuccessful. Averell died in 1986, survived by his third wife, Pamela Digby Churchill Heyward Harriman.[13]

E. Roland Harriman, the youngest of the Harriman children, married Gladys Fries and settled into a comfortable life, making his home at the big house on Echo Lake at Arden. Besides being a principal at W. A. Harriman and Co., and later at Brown Brothers Harriman and Co., working closely with Averell, his brother and best friend, he invested in diverse holdings over the years, many of them inspired by his interests. Roland invested in the start-up magazine *Time* in 1923 and *Newsweek* during its early years, which provided him with a taste of the media world. At one point he found himself the owner of a local newspaper based in Middletown, New York, that was about to go bankrupt. He was able to turn it around and ultimately sold it at a profit. The most famous of his projects, of course, was the Bear Mountain Bridge. He was responsible for creating the investment program that financed the construction, and he served as president of the Bear Mountain Hudson River Bridge Company that oversaw the project.[14]

He and his wife, Gladys, worked on behalf of the American Red Cross for much of their lives. From 1950 to 1953 he served as president, and from 1953 through to 1973, he was chairman of the board.

Roland died at home on February 17, 1978, at the age of eighty-two. He will be remembered for his philanthropic affiliations with the Boys Club of New York, the American Red Cross, and the New York Hospital–Cornell Medical Center.[15] And, of course, he will be remembered for his affiliation with the Bear Mountain Bridge.

Over the years the Harriman family donated most of their family estate to the state of New York. Landholdings they provided came with stipulations that they be developed as parks for public use or roadways, hospitals, and other projects intended to help meet community needs. In 1950 the family donated the Arden manor house to Columbia University for use as an extended campus location.[16]

The Harrimans' influence on the culture and history of the Hudson Valley was unmatched. Even today, descendants of E. H. Harriman provide continuing support to the valley through their affiliations with such organizations as the Palisades Interstate Park Commission and the Scenic Hudson and Open Space, to name a few; as of this writing, Averell Harriman's grandson David Mortimer only recently stepped down as chairman of the board for the PIPC. The family's influence continues to be deeply integrated into the valley.

Epilogue

Today the Hudson River valley remains very similar to the valley of the early twentieth century. Its quaint rural towns evoke a love of history and a respect for nature. Hiking trails remain untouched, and many city dwellers weekend in the hills just as they did one hundred years ago.

The Bear Mountain Bridge still stands majestic and elegant, just as it did in 1924, spanning the river, connecting Bear Mountain Park and Anthony's Nose. While no longer holding the distinction as the longest suspension bridge in the world, it remains a proud monument to the early twentieth century, a time when it opened a much-needed route for thousands of travelers, and its honored history is no less important.

Early in 1940, the Bear Mountain Bridge investors met and, after much discussion, decided to offer the state of New York an opportunity to purchase the bridge. Although the original charter stated that by 1954 (thirty years after the opening) the bridge would automatically revert to the state at no cost,[1] the investors, dissatisfied with the inadequate return that continued with no sign of improvement, decided that they did not want to wait another fourteen years to exit the investment. Since the bridge was generating just enough income to cover expenses, the state of New York decided that, at the right price, it would be beneficial for them to assume ownership. On September 25, 1940, the title was transferred from the Bear Mountain Hudson River Bridge Company to the state of New York in exchange for the sum of $2,275,000.[2]

Nine additional Hudson River crossings have been constructed since 1924, but the Bear Mountain Bridge maintains a prominent place as a monument to the builders and as a favorite among those people who admire bridges.

After ownership was transferred to the state in 1940, responsibility for management was assumed by the New York State Bridge Authority, an organization created under Franklin D. Roosevelt while he was governor of New York, in 1931. With the nation in the middle of the Great Depression, New York's tax revenue was scant. To address this situation, Governor Roosevelt proposed forming a new bridge authority to be funded by selling bonds secured with user-fee revenue rather than tax dollars, as was the norm at the time. By early 1932, the legislation passed, and the NYSBA was born. From this point forward, the construction, operation, and maintenance of Hudson River crossings between the Tappan Zee Bridge to the south and Castleton Bridge to the north would be self-supporting and would no longer present a burden to the taxpayer.

The New York State Bridge Authority is now operating in its ninety-fourth year as a public corporation governed by a five-member board appointed by the governor. It remains self-sufficient, obtaining its funding from authority bonds and bridge-user fees (tolls). Over the years, the NYSBA has exhibited a progressive attitude toward growth and development in the Hudson Valley, balancing economic expansion with conservation. The authority is currently responsible for the Bear Mountain Bridge, the Mid-Hudson Bridge, the Rip Van Winkle Bridge, the Newburgh-Beacon Bridge, the Kingston-Rhinecliff Bridge, and, as of 2010, the Walkway over the Hudson (the bridge structure only).[3]

Under the management of the New York State Bridge Authority, the Bear Mountain Bridge has maintained its place as an intricate part of the valley's culture. Maintenance is an important part of the NYSBA's program, and through their vigilance, they have ensured the bridge's journey through its second century as a productive part of New York's economy. From 1959 through 2002, the NYSBA invested approximately $35 million in the Bear Mountain Bridge for annual capital improvements, upgrades, and repairs. The largest expenditure was made in 1992 when $16.23 million was invested to upgrade the Bear Mountain Bridge toll plaza. The project included new tollbooths, road resurfacing, upgraded electrical lighting, and safety access to Bear Mountain Park.[4] As an indication of the bridge's continuing viability and growth, the following statistics bear witness: In 1983 a total of 3,802,678 vehicles crossed this monument of steel and grit, generating a net toll revenue of $936,855.60. By 2019, the number of vehicles reported crossing increased to 7,879,666 (more than double), and the revenue generated was $5,692,977 (an increase of more than 500 percent).

The bridge authority's success in maintaining and promoting the Bear Mountain Bridge is a feather in the cap of the state of New York, the Hudson Valley region, and the many people involved in its management. It is important to note that the NYSBA not only maintains the structures in its care but also supports the history of each. A museum dedicated to the Bear Mountain Bridge's history is located at the west-shore anchorage and can be visited by the public. Celebrations and commemorations of historical events related to all the bridges are planned and executed by the NYSBA with the cooperation and collaboration of the local towns and their historical societies.

The National Register of Historic Places is the official list of those historic places within America that have been deemed worthy of preservation; it was established in 1966, and the database is under the stewardship of the National Park Service. The Bear Mountain Bridge along with the original tollhouse located on the Peekskill approach road, Routes 6 and 202, were designated in 1982, and both are listed on the register.

November 26, 2024, was the centennial anniversary of the Bear Mountain Bridge. The New York State Bridge Authority held a historic, memorable celebration to mark this milestone. The NYSBA created a website specific to the celebration, providing history, information, event scheduling, and more, all relevant to the planned festivities.[5] Other historic organizations presented their testaments to the much-revered monument, and relevant information was published in the local newspapers and on the internet. I was privileged to receive an invitation to attend the festivities, and on a beautiful autumn day in November, I joined many others who appreciated the historical value of the legacies that are part of the Hudson River valley. Of course, there were representatives of the NYSBA in attendance as well as local dignitaries who spoke, honoring the bridge and its valued history. Two descendants of Roland and Averell Harriman were also in attendance. Roland's great-grandson and the spouse of Averell's grandson greeted many of the attendees and graciously spoke to the crowd, providing personal insight from the Harriman family. As a reflection of the original opening ceremonies that took place one hundred years earlier, several members of the West Point Band provided music for the ceremony. In deference to the 160 Studebakers that crossed the bridge on opening day in 1924, as well as to the many thousands of cars that have crossed the bridge over the past hundred years, there was a commemorative motorcade of antique cars for the 2024 centennial celebration. Classic cars, several from each decade representing the 1920s through to the present day, were invited to parade across the bridge as horns honked and bystanders cheered them on, all while the West Point Band continued to play. Following the outdoor festivities, just as Roland Harriman had provided the original attendees with a lovely luncheon, we too were invited to sit down to a meal at the Bear Mountain Inn, continuing the celebration amid testimonials and cordial conversation. Guests then adjourned to the historic Paramount Hudson Valley Theatre in Peekskill for the world premiere of a documentary film, *Bear Mountain Bridge: The First 100 Years*, produced by Historic Bridges of the Hudson Valley in conjunction with SDS Imagery and Tesseract Studios especially for the centennial festivities, and it was an appropriate finale to the celebration.[6]

While these events honored the bridge first and foremost, in doing so homage was paid to the men and women throughout history, and to the many who even today continue to dedicate themselves to preserving the environment and the beauty of the Hudson Valley—a locale that boasts some of the most beautiful natural assets in the country. These treasures might have been doomed to extinction if not for the dedicated citizens who, one hundred years ago, took matters into their own hands and prevented an outcome that they found intolerable. The many parts and pieces that came together because of their efforts created the world we enjoy today. Each, in their own way, has embedded a part of themselves into the Hudson Valley, and we are that much better because they did.

In writing this book, I have tried to keep alive the spirit of those persons who came before me and to chronicle the fascinating history of how the Bear Mountain Bridge was created and its environment was preserved.

This region rivals the most picturesque areas around the world, and the Bear Mountain Bridge stands today as a slender, elegant structure held aloft by a graceful sway of cabled steel set within the breathtaking scenery of the Hudson River valley—a reminder of the heritage bequeathed to us by those individuals who made it happen. The bridge may be made of *steel*, but it was their *grit* and determination that created its legacy.

Glossary

Notes

Bibliography

Index

Glossary

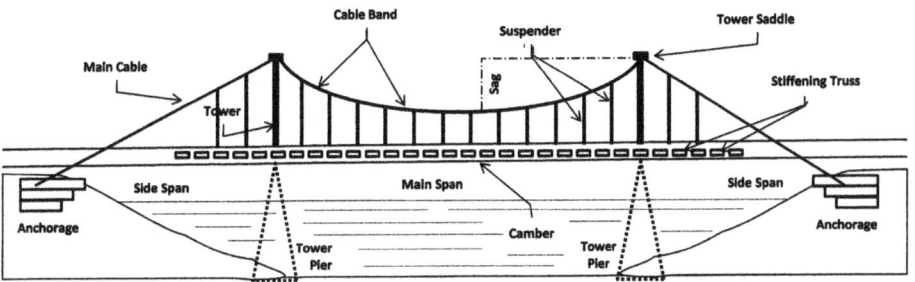

SUSPENSION BRIDGE TERMINOLOGY

In a suspension bridge, the roadway hangs from massive steel cables by shorter, vertical suspender cables. The long main cables are anchored at each end of the bridge in solid rock or concrete, and run up and over the bridge's two towers. The cables transfer the weight of the roadway and the bridge traffic to the towers. Anchored in the earth, the two towers do the job of supporting the bridge's weight.

33. Diagram of a suspension bridge with various parts identified. Courtesy of the Roebling Museum.

Abutment: A substructure that supports one end of the bridge and, at the same time, laterally supports the embankment that serves as an approach to the bridge.

Air compressor: A machine that provides compressed air through a receiver so that its expansion and the pressure thus created can be used as a power source.

Anchorage: In a suspension bridge, the anchorage is the location where the main cables that carry the load of the main span are anchored to the ground and encased in concrete.

Anchorage tunnel: A tunnel excavated to access the point of anchorage.

Anchor bars: An eyebar extending from the shoe of a span or tower into concrete or masonry used to hold down the span that it connects to and thus prevent uplift.

Anneal: A process used to reduce brittleness and increase the ductility of metal by heating it to a certain temperature and then cooling slowly in air or oil.

Approach span: The part of the bridge that carries traffic from the land to the main parts of the bridge.

Base castings: A steel or iron casting upon which the bridge shoe rests.

Beam: A member the prime function of which is to carry a transverse load.

Bond/surety bond: A contract that ensures the performance or payment of an obligation by one party to another.

Bonding company: An organization that provides surety bonds to contractors and businesses. The bonding company protects the party that requires the bond (project owner) from financial loss if the party that provides the bond (contractor) fails to meet its obligation.

Cable: A heavy rope, chain, or twisted wire rope suspending portions of a suspension bridge.

Catenary: A curve formed by a flexible, inextensible cable of uniform weight per unit of length, hung from two points (usually the towers) and supporting its own weight alone.

Cofferdam: A watertight enclosure around a work area that allows for the pumping out of water. It exposes the bed of a body of water to permit the construction of a pier or other hydraulic structures.

Creeper traveler: A small movable derrick running on a track on the upper chord of a truss. It usually has two booms.

Deck suspenders: A hanger used to suspend a floor/deck from a cable or truss or another object.

Deck system or floor system: The complete system of members forming a structure's floor or deck, which carries the floor and its load.

Derrick: An apparatus for lifting and moving heavy weights. It is similar to a crane but differs in having a boom, which corresponds to the jib of a crane, pivoted at the lower end so that it may take different inclinations.

Eyebars: A bar with an eye at one end or both ends.

Footbridge/footwalk: In bridge building, a temporary wooden walkway strung along the structure to allow workmen to access the bridge during construction.

Footbridge cable: A cable used to suspend a footbridge.
Foundation: A portion of a structure, usually below the surface of the ground, on which the structure rests and distributes pressure upon its support.
Hanger cable: A vertical suspender cable that transfers the load from the bridge deck to the main cables.
Parapet: A low wall or barrier placed on top of an abutment to hold back earth from encroaching on the end of the span.
Pier: A structure, usually made of masonry, used to transmit the loads from a bridge superstructure to the foundation.
Plaza: An open, flat space.
Progress payment: In construction contracts, it is a partial payment remitted usually monthly according to the work accomplished during that time period on the project.
Retainage: In construction contracts, an amount of 10 percent or 15 percent is deducted from the amount owed the contractor each month and held by the owner until the project is complete and the work is fully accepted, to guarantee that the contractor finishes the project and delivers all items agreed to in the contract.
Retaining wall: A wall built to sustain lateral pressure, such as an earth thrust.
Shop drawing: A drawing of a structure or a machine showing all parts, dimensions, and details with specific instructions related directly to one area of work.
Staging area: An area used to organize materials in anticipation of installing them.
Steam shovel: An excavating machine powered by steam.
Stiffening truss: A truss used in connection with a suspension cable to distribute the load over the length thereof.
Suspender: Suspender cables are short, vertical cables that support the main cables and distribute the loads evenly across the main cables.
Suspension cable: The main cable on a suspension bridge that is draped between the towers. Suspender cables are attached to it.
Suspension span: The distance between two intermediate supports (towers) on a suspension bridge.
Tower: A vertical structure consisting of two or more bents of framework connected by bracing.
Tower saddles: A block at the top of a tower of a suspension bridge over which passes the suspension cables.

Transverse bracing: Bracing that is perpendicular (or slightly inclined) to the centerline of the structure.

Viaduct: An extended bridge of many arches, piers, or columns that carries a road or railway over mainly dry ground.

Notes

1. You Are Cordially Invited to the Opening

1. "Opening Ceremony of Peekskill Bear Mountain Bridge," *Peekskill (NY) Highland Democrat*, Nov. 29, 1924.
2. "Cross Hudson Span in Autos This Week," *New York Times*, Nov. 23, 1924.
3. "Opening Ceremony."
4. *Construction of Parallel Wire Cables for Suspension Bridges*, pl. 67.
5. "Bridge Opening Draws Throng to Bear Mountain," *Middletown (NY) Daily Herald*, Nov. 2, 1924.
6. "Opening Ceremony."
7. "New Hudson Bridge Design Defended," *New York Times*, July 21, 1923.
8. E. Roland Harriman to C. A. Pugsley, Aug. 11, 1924, Harriman Collection, Orange County Historical Society Archives.
9. E. Roland Harriman to Hon. Alfred E. Smith, Nov. 7, 1924, Harriman Collection, Orange County Historical Society Archives.
10. "Opening Ceremony."
11. "Opening Ceremony."
12. "Opening Ceremony."
13. Documents belonging to E. Roland Harriman, guest list for the opening of the BMB, Harriman Collection, Orange County Historical Society Archives. Other guests (not listed in text) introduced at the ceremony included Holden D. Robinson, leading engineer on the suspension work; L. N. Gross, supervisor of placing the cables (as noted at the time, there was enough wire to stretch to San Francisco and back); W. E. Joyce, an assistant engineer on the cables; J. W. Caldwell, foreman; William Mallory, assistant superintendent of concrete; and J. W. Kemp, foreman.
14. H. G. Reynolds to E. R. Harriman, Dec. 6, 1924, Harriman Collection, Orange County Historical Society Archives.
15. "Opening Ceremony."
16. "Opening Ceremony."
17. I. Rossi to E. R. Harriman, Dec. 11, 1924, Harriman Collection, Orange County Historical Society Archives.

18. Various invitees to E. R. Harriman, Nov. 25–Dec. 1, 1924, Harriman Collection, Orange County Historical Society Archives.

19. "The Bear Mountain Bridge," excerpt from the New York Bridge Authority website, https://nysba.ny.gov/bridge/bear-mountain, accessed Dec. 17, 2024.

20. "Early Motor Use for Bear Mountain Bridge," *New York Times*, Oct. 26, 1924.

21. "Cross Hudson Span."

22. "The Bear Mountain Bridge," *Outlook*.

2. The Valley

1. William Harper Bennett, *Catholic Footsteps in Old New York: A Chronicle of Catholicity in the City of New York from 1524 to 1808*, 13.

2. Bennett, *Catholic Footsteps*, 10–11; Tom Lewis, *The Hudson: A History*, 41–42.

3. Lewis, *The Hudson: A History*, 45.

4. "Hudson River," in *Encyclopedia Britannica*, July 15, 2016, https://www.britannica.com/place/Hudson-River.

5. Lewis, *The Hudson: A History*, 89–92.

6. An American author, *American Husbandry: An Account of the Soil, Climate, Production, and Agriculture, of the British Colonies in North America and the West Indies*, 95–96.

7. Frances F. Dunwell, *The Hudson River Highlands*, 51–53.

8. Roderick Frazier Nash, *Wilderness and the American Mind*.

9. "Tugboats: Workhorses of the Hudson River."

10. Dunwell, *The Hudson River Highlands*, 9–10.

11. Dunwell, *The Hudson River Highlands*, 132.

12. "A Veteran Engineer's Death; John B. Jervis, Who Helped to Construct the Erie Canal," *New York Times*, Jan. 14, 1885.

13. Dunwell, *The Hudson River Highlands*, 133.

14. Dunwell, *The Hudson River Highlands*, 133.

15. Dunwell, *The Hudson River Highlands*, 132–34.

16. Donald F. Clark, "The Bridge That Never Was: Precursor of the Bear Mountain Hudson River Bridge."

17. "General Wellman Serrell Dead," *New York Times*, Apr. 26, 1906.

18. Clark, "Bridge That Never Was."

19. "The Hudson Suspension Bridge," *Harper's Weekly*, May 10, 1890.

20. Clark, "Bridge That Never Was."

21. Matthew Josephson, *The Robber Barons: The Great American Capitalists, 1861–1901*, 76–89.

22. Kenneth W. Maddox, "The Lure of the Country," 103.

23. US Fish and Wildlife Service, "History of the U.S. Fish and Wildlife Service."

24. Dunwell, *The Hudson River Highlands*, 140.

25. "The Grand Driveway," *Jersey City (NJ) News*, Jan. 25, 1895.

26. "To Save the Palisades, William Walter Phelps' Plan to Establish a Private Park," *Jersey City (NJ) News*, Feb. 6, 1894.

27. "State to Interfere: The Palisades May Be Turned into a Public Park," *Jersey City (NJ) News*, Sept. 24, 1894.

28. "To Save the Palisades, New Jersey People Planning Action," *New York Tribune*, Oct. 14, 1894.

29. "Bombarded by Palisades Rock," *Paterson (NJ) Evening News*, Sept. 2, 1899.

30. "To Save the Palisades, New Jersey People Planning Action."

31. "Stop Blasts on Palisade, Residents in New York State Claim Their Houses Are Damaged by Explosions," *Perth Amboy (NJ) Evening News*, Jan. 18, 1907.

32. "Save the Palisades, Efforts to Stop the Destruction of Hudson's Picturesque Scenery," *Jersey City (NJ) News*, Nov. 20, 1894.

33. "To Save the Palisades, New Jersey People Planning Action."

34. David Schuyler, *Sanctified Landscape: Writers, Artists, and the Hudson River Valley, 1820–1909*, 152.

35. H. Paul Jeffers, *The Bully Pulpit: A Teddy Roosevelt Book of Quotations*, 30.

36. Robert O. Binnewies, *Palisades: 100,000 Acres in 100 Years*, 37–39.

3. Wall Street and Railroads

1. Michelle P. Figliomeni, *E. H. Harriman at Arden Farms*.

2. George Kennan, *Railroad Tycoon: A Biography of E. H. Harriman*, chap. 3.

3. Maury Klein, *The Life and Legend of E. H. Harriman*, 68–69.

4. Klein, *Life and Legend of Harriman*, 29.

5. Donald E. Wolf, *Crossing the Hudson: Historic Bridges and Tunnels of the River*, 93.

6. Kennan, *Railroad Tycoon*, chap. 1.

7. Klein, *Life and Legend of Harriman*, 40–41.

8. Kennan, *Railroad Tycoon*, chap. 1.

9. Kennan, *Railroad Tycoon*, chap. 1.

10. Wolf, *Crossing the Hudson*, 94.

11. E. Roland Harriman, *I Reminisce*.

12. "Some Sidelights on the Character of Mr. Harriman Who Died Last Week," *New York Daily Tribune*, Sept. 12, 1909.

13. Klein, *Life and Legend of Harriman*, 45.

14. Wolf, *Crossing the Hudson*, 94–95.

15. Kennan, *Railroad Tycoon*, chap. 3.

16. Klein, *Life and Legend of Harriman*, 95.

17. Figliomeni, *Harriman at Arden Farms*.

18. Klein, *Life and Legend of Harriman*, 69.

19. Figliomeni, *Harriman at Arden Farms*.
20. Wolf, *Crossing the Hudson*, 96.
21. Figliomeni, *Harriman at Arden Farms*.
22. Figliomeni, *Harriman at Arden Farms*.
23. Stephen Birmingham, *Our Crowd: The Great Jewish Families of New York*, 170–73, 182, 201.
24. Birmingham, *Our Crowd*, 201.
25. Birmingham, *Our Crowd*, 170–73.
26. Klein, *Life and Legend of Harriman*, 64–65.
27. Birmingham, *Our Crowd*, 203, 204–8, 169, 204–8.
28. "Harriman Builder of Railroad Empire," *New York Times*, Sept. 10, 1909.
29. "Northern Pacific Peace," *New York Daily Tribune*, July 18, 1909.
30. Klein, *Life and Legend of Harriman*, chap. 25.

4. The Right Man for the Job

1. Binnewies, *Palisades*, 39, 41.
2. John A. Garraty, *Right Hand Man: The Life of George W. Perkins*, 6–10.
3. Garraty, *Right Hand Man*, 15–16.
4. Garraty, *Right Hand Man*, 19–22.
5. "Beers Is Down and Out," *New York Times*, Feb. 9, 1892.
6. Garraty, *Right Hand Man*, 35–42.
7. Michael Middleton Dwyer, ed., *Great Houses of the Hudson River*.
8. Garraty, *Right Hand Man*, 59–62, 65–72.
9. Garraty, *Right Hand Man*, 83–84.
10. Binnewies, *Palisades*, 11–15.
11. "Preserve the Palisades, New Jersey Leaders Say the State Should Take Action," *New York Times*, Oct. 10, 1895.
12. Commissioners of the Palisades Interstate Park, *The Palisades Interstate Park, 1900–1929: A History of Its Origin and Development*.
13. Cecilia Gaines Holland, "The Saving of the Palisades," *New Jersey Bulletin*, a publication of the New Jersey State Federation of Women's Clubs, Nov. 1930, New Jersey State Federation of Women's Clubs Historical Archives.
14. "The Question of the Palisades," *New York Sun*, Dec. 3, 1897.

5. The Original Jersey Girls

1. New Jersey State Federation of Women's Clubs (hereafter cited as NJSFWC) Historical Archives.
2. "Vassar Class of 1873, 50th Reunion Notes," contemporary notes, 201, Historical Archives of the Women's Club of Englewood, courtesy of club historian Janine B. McKee, 2021.

3. "Before Women Clubbed," *Jersey City (NJ) News*, May 14, 1897.

4. Binnewies, *Palisades*, 9.

5. Palisades Interstate Park Commission, *60 Years of Park Cooperation*.

6. Binnewies, *Palisades*, 10, 11.

7. Cecilia Gaines Holland, "The Saving of the Palisades," *New Jersey Bulletin*, a publication of the NJSFWC, Nov. 1930, NJSFWC Historical Archives.

8. Binnewies, *Palisades*, 11–12.

9. "To Save the Palisades—Inspect Hills from a Yacht," *New York World*, Sept. 28, 1897.

10. "To Save the Palisades—Inspect Hills."

11. "To Save the Palisades, Action Revoking Contractors' Lease to Be Reviewed in New Jersey," *New York Times*, Nov. 25, 1897.

12. "To Save the Palisades, Dr. West Suggests That the Brooklyn Institute Take a Hand," *Brooklyn Daily Eagle*, Nov. 17, 1897.

13. "Governor's Message," *Gloucester County (NJ) Democrat*, Jan. 11, 1900.

14. Binnewies, *Palisades*, 12.

15. Binnewies, *Palisades*, 14–15.

16. Binnewies, *Palisades*, 15.

17. Holland, "Saving of the Palisades."

18. "The Strenuous Efforts to Save the Palisades," *Newark Daily Advertiser*, Feb. 26, 1901.

19. "Second President, 1896–1898; Dean of Past Presidents, 1928–1943," NJSFWC Historical Archives.

20. "In Appreciation," remembrance written by a club member of the NJSFWC in tribute to C. G. Holland, president emeritus, NJSFWC Historical Archives.

21. Biography of Cecilia Gaines Holland, written by a club member, NJSFWC Historical Archives.

22. Palisades Interstate Park Commission, *60 Years of Park Cooperation*.

6. A Perfect Candidate

1. John A. McCall to G. W. Perkins (telegram), Mar. 29, 1900, George W. Perkins Sr. Personal Papers, Rare Book & Manuscript Library, Columbia University.

2. Binnewies, *Palisades*, 15–16.

3. Binnewies, *Palisades*, 16.

4. Commissioners of the Palisades Interstate Park, *Palisades Interstate Park*.

5. Binnewies, *Palisades*, 18.

6. Arthur Carlyle Mack, *The Palisades of the Hudson: Their Formation, Tradition, Romance, Historical Associations, Natural Wonders and Preservation*, 43.

7. "Palisades Blasting Stops," *New York Tribune*, Dec. 29, 1900.

8. Garraty, *Right Hand Man*, 84.

9. Elizabeth B. Vermilye to George W. Perkins Sr., July 1900, Perkins Sr. Personal Papers, Rare Book & Manuscript Library, Columbia University.

10. Binnewies, *Palisades*, 16.

11. George W. Perkins to Elizabeth B. Vermilye, July 26, 1900, Perkins Sr. Personal Papers, Rare Book & Manuscript Library, Columbia University.

12. Elizabeth B. Vermilye to G. W. Perkins, July 28, 1900, Perkins Sr. Personal Papers, Rare Book & Manuscript Library, Columbia University.

13. John Leonard, ed., *Woman's Who's Who in America*, 838.

14. "Bible Lecture Interests Women," *Perth Amboy (NJ) Evening News*, Mar. 24, 1923.

15. Garraty, *Right Hand Man*, 85–86.

16. "Palisades Blasting Stops," *New York Tribune*, Dec. 29, 1900.

7. A Partnership Is Formed

1. Binnewies, *Palisades*, 15–16.

2. Erica Wagner, *Chief Engineer: Washington Roebling, the Man Who Built the Brooklyn Bridge*, xviii–xix.

3. Binnewies, *Palisades*, 20.

4. Katharine J. Sauzade to Cecilia G. Holland, Mar. 2, 1901, New Jersey State Federation of Women's Clubs (hereafter cited as NJSFWC) Historical Archives.

5. Figliomeni, *Harriman at Arden Farms*.

6. Wolf, *Crossing the Hudson*, 96.

7. Figliomeni, *Harriman at Arden Farms*.

8. Figliomeni, *Harriman at Arden Farms*.

9. "Tower Hill's Beauties," *New York Daily Tribune*, June 27, 1909.

10. Binnewies, *Palisades*, 30–31.

11. *Seventh Annual Report of the Commissioners of the Palisades Interstate Park Commission (for Year 1906)* (J. B. Lyon, Printers, 1907), Palisades Interstate Park Commission (hereafter cited as PIPC) Historical Archives.

12. Palisades Interstate Park Commission, *60 Years of Park Cooperation*.

13. Binnewies, *Palisades*, 29.

14. Receipt issued by the PIPC to the NJSFWC for $3,039.39, dated May 23, 1908, deposit provided for the construction of a memorial to the NJSFWC, PIPC Historical Archives, New York.

15. J. DuPratt White to Mrs. Henry H. Dawson, Mar. 4, 1908, PIPC Historical Archives.

16. "Bergen County Historical Society Formed There by Residents," *New York Tribune*, Apr. 13, 1902.

17. Chelsea Gibson, *Biographical Sketch of Adaline Wheelock Sterling*.

18. Gibson, *Adaline Wheelock Sterling*.

8. Progress and Development

1. *Twentieth Annual Report of the Commissioners of the Palisades Interstate Park Commission* (J. B. Lyon, Printers, 1920), Palisades Interstate Park Commission (hereafter cited as PIPC) Historical Archives.
2. *Seventh Annual Report of the Commissioners of the Palisades Interstate Park Commission (for the Year 1906)* (J. B. Lyon, Printers, 1907), PIPC Historical Archives.
3. Dunwell, *The Hudson River Highlands*, 144–49.
4. Dunwell, *The Hudson River Highlands*, 167.
5. Dunwell, *The Hudson River Highlands*, 169, 175.
6. Dunwell, *The Hudson River Highlands*, 179–83.
7. Palisades Interstate Park Commission, *60 Years of Park Cooperation*.
8. Figliomeni, *Harriman at Arden Farms*.

9. The Devil's Horse Race

1. Figliomeni, *Harriman at Arden Farms*.
2. Binnewies, *Palisades*, 39.
3. Binnewies, *Palisades*, 35, 235.
4. Binnewies, *Palisades*, 26.
5. Binnewies, *Palisades*, 75.
6. Dr. E. L. Partridge, "A National Park on the Hudson."
7. Binnewies, *Palisades*, 36–37.
8. Wolf, *Crossing the Hudson*, 97.
9. Klein, *Life and Legend of Harriman*, 431–32.
10. Robert H. Fuller, Esq., secretary for Governor Hughes, and Mr. Thomas Price, secretary for E. H. Harriman, Feb. 6–June 25, 1909, Harriman Collection, Orange County Historical Society Archives.
11. Figliomeni, *Harriman at Arden Farms*.
12. Fuller and Price, Feb. 6–June 25, 1909.
13. Klein, *Life and Legend of Harriman*, 440–41.
14. "Harriman Dead; News Delayed," *New York Times*, Sept. 10, 1909.
15. Klein, *Life and Legend of Harriman*, 444.
16. "Minutes of a Special Meeting of the Commissioners of the Palisades Interstate Park, New York," Dec. 16, 1909, Palisades Interstate Park Commission (hereafter cited as PIPC) Historical Archives.
17. Dunwell, *The Hudson River Highlands*, 164.
18. "Minutes of a Special Meeting of the Commissioners of the Palisades Interstate Park, New York," Dec. 23, 1909, PIPC Historical Archives.
19. "Minutes of a Special Meeting."
20. Binnewies, *Palisades*, 51–54.

21. "Palisades Park Plan Now Up to Voters," *New York Tribune*, Oct. 30, 1910.
22. As quoted in Binnewies, *Palisades*, 54.

10. The Chief Engineer

1. "Major Welch Dies; Builder of Parks," *New York Times*, May 5, 1941.
2. Ruby M. Jolliffe, *Biography of William A. Welch*, on behalf of PIPC for the National Conference on State Parks, May 8, 1946, Palisades Interstate Park Commission (hereafter cited as PIPC) Historical Archives, New York.
3. Binnewies, *Palisades*, 79.
4. Jolliffe, *Biography of William A. Welch*.
5. Wendy Welch, daughter of William A. Welch Jr., granddaughter of Major William A. Welch, interview by the author.
6. William A. Welch and Camille Beall married in 1904, per Marriage Records of the State of New Jersey, https://ancestry.com, accessed Nov. 2024.
7. Ship Manifest, SS *Obidense*, sailing from Para, Brazil, Mar. 22, 1908, https://ancestry.com, accessed Nov. 2024.
8. US Federal Census records, 1910, Borough of Queens, State of New York, https://ancestry.com, accessed Nov. 2024.
9. "Bear Mountain Inn."
10. Binnewies, *Palisades*, 214.
11. Binnewies, *Palisades*, 87, 194.
12. Binnewies, *Palisades*, 95.
13. Binnewies, *Palisades*, 95.
14. *Sixteenth Annual Report of the Commissioners of the Palisades Interstate Park Commission* (J. B. Lyon, Printers, 1916), PIPC Historical Archives.
15. "Proceedings of the National Parks Conference," Department of the Interior, National Parks Service, Jan. 4, 1917, PIPC Historical Archives.
16. Meade C. Dobson, "League of Walkers Proposed to Unite Lovers of Outdoors," *New York Evening Post*, Sept. 24, 1920.
17. Raymond H. Torrey, Frank Place Jr., and Robert L. Dickinson, *New York Walk Book*.
18. Benton MacKaye, "An Appalachian Trail, a Project in Regional Planning."
19. William H. Carr, "Signs along the Trail."
20. Statistics courtesy of the Appalachian Trail Conservancy, Harpers Ferry, WV, https://appalachiantrail.org/explore/hike-the-a-t/thru-hiking/faqs/.
21. *Twentieth Annual Report of the Commissioners of the Palisades Interstate Park Commission* (J. B. Lyon, Printers, 1920), PIPC Historical Archives.
22. Binnewies, *Palisades*, 117.
23. Binnewies, *Palisades*, 99–101, 147.

11. Like Father, Like Son

1. Garraty, *Right Hand Man*, 375.
2. Draft copy of a bio titled "G. W. Perkins, Jr.," Perkins Sr. Personal Papers, Rare Book & Manuscript Library, Columbia University.
3. Ticket stubs and memorabilia, Perkins Sr. Personal Papers, Rare Book & Manuscript Library, Columbia University.
4. "George W. Perkins, Jr., Weds at Princeton," *New York Sun*, June 20, 1917.
5. "Perkins Leads Yaphank Squad," *New York Globe*, Sept. 8, 1917.
6. "Drafted Men Selected for Officers School," *New York Tribune*, Jan. 3, 1918.
7. Garraty, *Right Hand Man*, 376–77.
8. "Mrs. G. W. Perkins, Jr., Dies," *New York Tribune*, Oct. 8, 1918.
9. Garraty, *Right Hand Man*, 377.
10. Garraty, *Right Hand Man*, 378–79, 382.
11. G. W. Perkins Sr. to William A. Welch, Jan. 28, 1919, Palisades Interstate Park Commission (hereafter cited as PIPC) Historical Archives.
12. "G. W. Perkins, Jr., Takes 'Y' Post," *New York Sun*, Apr. 26, 1919.
13. "4,000 College Graduates to Study Civic Problems," *New York Tribune*, Dec. 5, 1920.
14. "Committee Formed Here to Aid Poles in Fight with 'Reds,'" *New York Tribune*, Jan. 4, 1920.
15. Garraty, *Right Hand Man*, 389.
16. "Geo. W. Perkins Dies, Martyr to Public Service," *New York Herald*, June 19, 1920.
17. "Achievements of George W. Perkins."
18. W. A. Welch to Dr. George F. Kunz, June 19, 1920, PIPC Historical Archives.
19. "George W. Perkins, Jr., to Resign," *New York Times*, Mar. 25, 1922.
20. "G. W. Perkins, Jr., Weds Miss Merck in Grace Church," *New York Herald*, Dec. 18, 1921.
21. "Who's Who in the Western Hemisphere 1942," George W. Perkins Jr. Personal Papers, Rare Book & Manuscript Library, Columbia University.
22. "George W. Perkins, Jr., Gets State Office," *New York Herald*, Mar. 15, 1922.

12. We Must Not Forget

1. S. E. Demarest to J. D. White, Mar. 2, 1915, Palisades Interstate Park Commission (hereafter cited as PIPC) Historical Archives.
2. I. W. Dawson to White, Mar. 14, 1921, PIPC Historical Archives.
3. White to Dawson, Mar. 15, 1921, PIPC Historical Archives.
4. Dawson to White, Mar. 20, 1921, PIPC Historical Archives.

5. Wolf, *Crossing the Hudson*, 68.

6. Binnewies, *Palisades*, 36–50.

7. David E. Kyvig, *Daily Life in the United States, 1920–1940: How Americans Lived through the Roaring Twenties and the Great Depression*, 31.

8. "Hudson River Bridge at Bear Mountain: A Solution of the Motor Traffic Problem across the Upper Hudson River."

9. "Hudson River Bridge at Bear Mountain."

10. Henry Petroski, *Engineers of Dreams: Great Bridge Builders and the Spanning of America*, chap. 4.

11. "Bridging the Hudson," *New York Times*, Apr. 20, 1923.

12. "New Scenic Motor Highway in Highlands of the Hudson," *New York Times*, May 6, 1923.

13. A Man with a Plan

1. E. Roland Harriman to Frederick Tench, Jan. 12, 1922, Harriman Collection, Orange County Historical Society Archives.

2. Tench to Harriman, Jan. 17, 1922, Harriman Collection, Orange County Historical Society Archives.

3. Harriman, *I Reminisce*, 5–6.

4. John Muir, *Tribute to Edward Henry Harriman*, 10.

5. Klein, *Life and Legend of Harriman*, 187.

6. Harriman, *I Reminisce*, 5, 6.

7. Harriman, *I Reminisce*.

8. Klein, *Life and Legend of Harriman*, 195.

9. "History of Harriman State Park of Idaho."

10. Rudy Abramson, *Spanning the Century: The Life of W. Averell Harriman, 1891–1986*, 22.

11. Abramson, *Spanning the Century*, 106–7, 261.

12. Abramson, *Spanning the Century*, 260.

13. Abramson, *Spanning the Century*.

14. Abramson, *Spanning the Century*, 86–87, 116.

15. Harriman, *I Reminisce*.

16. Harriman to Tench, Jan. 27, 1922, Harriman Collection, Orange County Historical Society Archives.

17. Binnewies, *Palisades*, 36–41.

18. Harriman to Edward L. Partridge, Jan. 27, 1922, Harriman Collection, Orange County Historical Society Archives.

19. Tench to Harriman, Jan. 28, 1922, Harriman Collection, Orange County Historical Society Archives.

14. Baird

1. *Shop Journal of the Phoenix Bridge Co.* 701 (1896–1900), Phoenix Bridge Company Published Collections.
2. Thomas R. Winpenny, *Without Fitting, Filing, or Chipping: Illustrated History of the Phoenix Bridge Company*, x, 1.
3. Records of the Institution of Civil Engineers, membership documents, 1929.
4. Howard Carter Baird, enlisted June 16, 1898, Pennsylvania, US, National Guard Veterans Card Files, 1867–1921.
5. Patrick McSherry, "The History of the 3rd U.S. Volunteer Engineers."
6. US Federal Census records, June 1, 1900, Phoenixville Borough, Chester County, PA, https://www.ancestry.com/discovery, accessed Nov. 2024.
7. *Shop Journal of the Phoenix Bridge Co.*
8. Records of the Institution of Civil Engineers.
9. Records of the Institution of Civil Engineers.
10. Howard Carter Baird [author assumed], "Bear Mountain Hudson River Bridge." This document, in its entirety, will be available to readers interested on the author's website.
11. "Memoir of Henry Wilson Hodge."
12. Wolf, *Crossing the Hudson*, 99–100.

15. Bear Mountain Hudson River Bridge Company

1. "Bill Authorizes Huge Suspension Span over Hudson," *Brooklyn Daily Eagle*, Feb. 8, 1922.
2. "Continue to See Hope for Bridge Here," *Poughkeepsie (NY) Eagle-News*, Mar. 16, 1922.
3. "Bridge to Bear Mountain from Peekskill Is Passed; Po'keepsie Project 'Dies,'" *Yonkers (NY) Herald Statesman*, Mar. 18, 1922.
4. "Hudson River Bridge at Bear Mountain."
5. "Propose Bridge to Cross the Hudson River," *Bernardsville (NJ) News*, Mar. 23, 1922.
6. "Propose Bridge to Cross the Hudson River."
7. "Governor Signs Bill for Hudson Bridge," *New York Times*, Apr. 1, 1922.
8. "To Finance Bridge at Bear Mountain," *New York Times*, Apr. 19, 1923.
9. "Act to Regulate the Construction of Bridges over Navigable Waters" (33 USC, 491), 3, US law originally approved and enacted Mar. 23, 1906.
10. G. A. Ellis to E. R. Harriman, Nov. 8, 1922, Harriman Collection, Orange County Historical Society Archives.
11. Ellis to Harriman, Nov. 8, 1922, Harriman Collection, Orange County Historical Society Archives.

12. Ellis to Harriman (correspondence attachment), "Analysis of Charter," Nov. 8, 1922, Harriman Collection, Orange County Historical Society Archives.

16. Terry & Tench

1. "Men Who Really Built the Subway," *New York Times*, Nov. 4, 1904.
2. Tench family history notes, provided by Margaret Tench Hamilton Crothers, granddaughter of Frederick Tench.
3. Eva Elliott Tolan, "Reminders of the Past: The Tench Family," *Niagara Falls Review*, Feb. 9, 23, 1963.
4. "Edward F. Terry Obituary."
5. Kennan, *Railroad Tycoon*, chap. 5.
6. "Men Who Really Built the Subway."
7. "Men Who Really Built the Subway."
8. Society news, *Philadelphia Times*, Nov. 5, 1899, 28.
9. Tench family history notes.
10. Wolf, *Crossing the Hudson*, 243.
11. "To Develop NJ Meadows," *New York Sun*, Apr. 12, 1914.
12. "Men Who Really Built the Subway."
13. "The History of Ironworkers Local 361."
14. Arnold Manoff and Chris Thorsten, "Random Interviews of Union Ironworkers Regarding Job Issues and Coworkers."
15. "History of Ironworkers Local 361."
16. "Guard Peekskill Bridge," *Brooklyn Daily Eagle*, June 14, 1924.
17. "Union Iron Workers Locked Out on Bridge, They Claim; State Troopers Guarding It," *Peekskill (NY) Evening Star*, June 16, 1924.
18. Patent No. US900951A, status expired as of Oct. 13, 1925, US Patent and Trademark Office, data provided by IFI Claims, Patent Services, https://patents.google.com/patent/US900951A/en.
19. "Terry & Tench, the Famous Contractors, Invite You to Join Them . . . ," *Brooklyn Daily Eagle*, Mar. 31, 1910.
20. "Men Who Really Built the Subway."
21. "Place Big Bridge Span," *New York Tribune*, June 17, 1906.
22. "Plans for Construction of Twenty U.S. Vessels," *Washington (DC) Evening Star*, June 22, 1917.
23. "Construction of the Lincoln Memorial."
24. "Construction of the Lincoln Memorial."
25. Harriman, *I Reminisce*, 60.
26. Abramson, *Spanning the Century*, 210, 237.
27. Klein, *Life and Legend of Harriman*, 446–47.

28. Abramson, *Spanning the Century*, 195.
29. Harriman, *I Reminisce*.

17. A Difference of Opinion

1. Ed (associate of G. W. Perkins Jr.) of G. Amsinck and Company, Inc., to George W. Perkins Jr., May 4, 1922, Harriman Collection, Orange County Historical Society Archives.
2. Ed to Perkins Jr., May 4, 1922, Harriman Collection, Orange County Historical Society Archives.
3. Ed to Perkins Jr., May 4, 1922, Harriman Collection, Orange County Historical Society Archives.
4. Ed to Perkins Jr., May 4, 1922, Harriman Collection, Orange County Historical Society Archives.
5. E. R. Harriman to Frederick Tench, Jan. 17, 1923, Harriman Collection, Orange County Historical Society Archives.
6. Harriman to General Cornelius Vanderbilt, Jan. 31, 1923, Harriman Collection, Orange County Historical Society Archives.
7. Document: "Estimate of Cost on Basis of Signing Contract 1/1/23 and Completing Bridge and Approaches by 1/1/25," Nov. 6, 1922, submitted by Terry & Tench Co., Inc., to Bear Mountain Hudson River Bridge Company, Harriman Collection, Orange County Historical Society Archives.
8. G. A. Ellis to Harriman, July 3, 1923, and attachment: Ellis's draft of a letter to be sent from Harriman to Tench regarding contract billing, Harriman Collection, Orange County Historical Society Archives.
9. Harriman, *I Reminisce*, 57–58.
10. Harriman to E. J. Berwind, Feb. 6, 1923, Harriman Collection, Orange County Historical Society Archives.
11. Harriman to Henry R. Towne, Feb. 21, 1923, Harriman Collection, Orange County Historical Society Archives.
12. "Bonds for Bridge at Bear Mountain," *New York Times*, Apr. 20, 1923.
13. Ellis to Harriman, Mar. 22, 1923, Harriman Collection, Orange County Historical Society Archives.
14. Wilson Fitch Smith, "Bridging the Hudson River at Bear Mountain."
15. Smith, "Bridging the Hudson."
16. Harriman to Bear Mountain Hudson River Bridge Company, Mar. 24, 1923, Harriman Collection, Orange County Historical Society Archives.
17. T. W. Farnum to J. W. Wear, Apr. 2, 1923, Harriman Collection, Orange County Historical Society Archives.
18. Wear to Harriman, Apr. 3, 1923, Harriman Collection, Orange County Historical Society Archives.

19. Mortgage and deed of trust, Apr. 1, 1923, Harriman Collection, Orange County Historical Society Archives.

20. Notice of annual meeting, Apr. 10, 1923, Harriman Collection, Orange County Historical Society Archives.

18. Construction Begins

1. "Bear Mountain Bridge," *Peekskill (NY) Highland Democrat*, Apr. 7, 1923.
2. "Excavations for Bear Mountain Bridge Towers Under Way," *Scarsdale (NY) Inquirer*, July 21, 1923.
3. Lawrence Abbott, "Comments on the Bear Mountain Bridge."
4. "Design of Bear Mountain Bridge Attacked as Ugly."
5. "New Hudson Bridge Design Defended," *New York Times*, July 21, 1923.
6. "Bridge Design and Good Taste."
7. Baird [author assumed], "Bear Mountain Hudson River Bridge."
8. Sharon Reier, *The Bridges of New York*, 35.
9. "Othmar Hermann Ammann"; Frank Griggs Jr., "Great Achievements: Othmar H. Ammann," 44.
10. W. F. Smith to E. R. Harriman, "Bear Mountain Hudson River Bridge Company Monthly Progress Report," June 30, 1923, Harriman Collection, Orange County Historical Society Archives.
11. "Excavations for Bear Mountain Bridge Towers Under Way."
12. W. F. Smith to Harriman, July 14, 1923, Harriman Collection, Orange County Historical Society Archives.
13. Smith to Harriman, July 14, 1923, Harriman Collection, Orange County Historical Society Archives.
14. Baird [author assumed], "Bear Mountain Hudson River Bridge."
15. W. F. Smith to Harriman, "Bear Mountain Hudson River Bridge Company Monthly Progress Report," July 31, 1923, Harriman Collection, Orange County Historical Society Archives.
16. Frederick Tench to E. R. Harriman, July 6, 1923, Harriman Collection, Orange County Historical Society Archives.
17. Harriman to Tench, July 6, 1923, Harriman Collection, Orange County Historical Society Archives.
18. W. F. Smith to Harriman, "Bear Mountain Hudson River Bridge Company Monthly Progress Report," Aug. 31, 1923, Harriman Collection, Orange County Historical Society Archives.
19. Baird [author assumed], "Bear Mountain Hudson River Bridge."
20. W. F. Smith to Harriman, "Bear Mountain Hudson River Bridge Company Monthly Progress Report," Oct. 1, 1923, Harriman Collection, Orange County Historical Society Archives.

21. Smith to Harriman, "Bear Mountain Hudson River Bridge Company Monthly Progress Report," Oct. 1, 1923, Harriman Collection, Orange County Historical Society Archives.

22. Agreement between Rowland Hazard and Helen Campbell and BMHRBC and the PIPC, Sept. 25, 1923, recorded in Westchester County Registrar's office Oct. 11, 1923.

23. W. F. Smith to Harriman, "Bear Mountain Hudson River Bridge Company Monthly Progress Report," Nov. 1, 1923, Harriman Collection, Orange County Historical Society Archives.

24. W. F. Smith to Harriman, "Bear Mountain Hudson River Bridge Company Monthly Progress Report," Dec. 1, 1923, Harriman Collection, Orange County Historical Society Archives.

25. Smith to Harriman, "Bear Mountain Hudson River Bridge Company Monthly Progress Report," Dec. 1, 1923, Harriman Collection, Orange County Historical Society Archives.

26. "Contract of Sale, between Franklin Couch, et al., Clifford Couch, et al., and the BMHR Bridge Co.," Dec. 26, 1923, County of Westchester, State of New York.

27. Smith to Harriman, "Bear Mountain Hudson River Bridge Company Monthly Progress Report," Dec. 1, 1923, Harriman Collection, Orange County Historical Society Archives.

28. Smith to Harriman, "Bear Mountain Hudson River Bridge Company Monthly Progress Report," Dec. 1, 1923, Harriman Collection, Orange County Historical Society Archives.

29. W. F. Smith to Harriman, "Bear Mountain Hudson River Bridge Company Monthly Progress Report," Jan. 2, 1924, Harriman Collection, Orange County Historical Society Archives.

30. Tony Tekaroniake Evans, "How Mohawk 'Skywalkers' Helped Build New York City's Tallest Skyscrapers."

31. "Towers of First Suspension Bridge over Hudson River Completed."

32. Baird [author assumed], "Bear Mountain Hudson River Bridge."

33. Baird [author assumed], "Bear Mountain Hudson River Bridge."

34. "Bear Mountain Bridge Towers in Course of Construction," *Peekskill (NY) Evening Star*, Jan. 7, 1924.

35. Reier, *Bridges of New York*, 29–34.

36. W. F. Smith to E. R. Harriman, "Bear Mountain Hudson River Bridge Company Monthly Progress Report," Feb. 1, 1924, Harriman Collection, Orange County Historical Society Archives.

37. Harriman to L. Sells, Feb. 15, 1924, Harriman Collection, Orange County Historical Society Archives.

38. Harriman to A. H. Smith, Feb. 15, 1924, Harriman Collection, Orange County Historical Society Archives.

39. Smith to Harriman, "Bear Mountain Hudson River Bridge Company Monthly Progress Report," Feb. 1, 1924, Harriman Collection, Orange County Historical Society Archives.

40. W. F. Smith to Harriman, "Bear Mountain Hudson River Bridge Company Monthly Progress Report," Mar. 1, 1924, Harriman Collection, Orange County Historical Society Archives.

41. "Towers of First Suspension Bridge over Hudson River Completed."

42. Smith to Harriman, "Bear Mountain Hudson River Bridge Company Monthly Progress Report," Mar. 1, 1924, Harriman Collection, Orange County Historical Society Archives.

43. W. F. Smith to Harriman, "Bear Mountain Hudson River Bridge Company Monthly Progress Report," Apr. 1, 1924, Harriman Collection, Orange County Historical Society Archives.

44. Smith to Harriman, "Bear Mountain Hudson River Bridge Company Monthly Progress Report," Apr. 1, 1924, Harriman Collection, Orange County Historical Society Archives.

45. "Bridge Blast a Railroad Delay," *Peekskill (NY) Evening Star*, Apr. 11, 1924.

46. W. F. Smith to Harriman, "Bear Mountain Hudson River Bridge Company Monthly Progress Report," May 1, 1924, Harriman Collection, Orange County Historical Society Archives.

47. "Footways Span Hudson on New Bridge Cables," *Peekskill (NY) Evening Star*, May 21, 1924.

48. *Construction of Parallel Wire Cables*, 7–9.

49. Baird [author assumed], "Bear Mountain Hudson River Bridge."

50. Smith to Harriman, "Bear Mountain Hudson River Bridge Company Monthly Progress Report," May 1, 1924, Harriman Collection, Orange County Historical Society Archives.

51. W. F. Smith to Harriman, "Bear Mountain Hudson River Bridge Company Monthly Progress Report," June 1, 1924, Harriman Collection, Orange County Historical Society Archives.

19. Innovations

1. *Construction of Parallel Wire Cables*, 6.

2. *Construction of Parallel Wire Cables*, 7–9.

3. "Bear Mountain Bridge—75th Anniversary."

4. George W. Perkins Jr. to J. DuPratt White, July 27, 1928, Harriman Collection, Orange County Historical Society Archives.

5. Perkins Jr. to White, July 27, 1928, Harriman Collection, Orange County Historical Society Archives.

6. "Edward F. Terry Obituary," 914.

7. W. F. Smith to E. R. Harriman, "Bear Mountain Hudson River Bridge Company Monthly Progress Report," July 1, 1924, Harriman Collection, Orange County Historical Society Archives.

8. *Construction of Parallel Wire Cables.*

9. W. F. Smith to Harriman, "Bear Mountain Hudson River Bridge Company Monthly Progress Report," Aug. 1, 1924, Harriman Collection, Orange County Historical Society Archives.

10. Henry J. Stanton, "View: Bear Mountain Bridge's Secret History," *LOHUD*, Nov. 27, 2014, https://www.lohud.com/story/opinion/contributor/2014/11/27/bear-mountain-bridge-secret-history/19394987/.

11. Baird [author assumed], "Bear Mountain Hudson River Bridge."

12. W. F. Smith to Harriman, "Bear Mountain Hudson River Bridge Company Monthly Progress Report," Sept. 2, 1924, Harriman Collection, Orange County Historical Society Archives.

13. Baird [author assumed], "Bear Mountain Hudson River Bridge."

14. Baird [author assumed], "Bear Mountain Hudson River Bridge."

15. W. F. Smith to Harriman, "Bear Mountain Hudson River Bridge Company Monthly Progress Report," Oct. 1, 1924, Harriman Collection, Orange County Historical Society Archives.

16. "Worker on Bridge Is Ground to Pieces by Central Train Today," *Peekskill (NY) Evening Star*, Apr. 15, 1924.

17. "Another Bridge Man Is Cut to Pieces on the Railroad Sunday," *Peekskill (NY) Evening Star*, Apr. 21, 1924.

18. "Finish Bridge Span, Longest in World," *New York Times*, Oct. 9, 1924.

19. "Drove Nail into Head by a 40-Foot Fall," *Brooklyn Citizen*, Aug. 13, 1907.

20. "Bridge Worker Drops 175 Feet into River, Lives to Tell Tale," *Elmira (NY) Star-Gazette*, Oct. 24, 1930.

21. "The Man on the Skyscraper," *Brooklyn Daily Eagle*, June 16, 1910.

22. "New Bridge Ablaze," *New York Sun*, Nov. 11, 1902.

23. "Man on the Skyscraper."

24. W. F. Smith to Harriman, "Bear Mountain Hudson River Bridge Company Monthly Progress Report," Nov. 1, 1924, Harriman Collection, Orange County Historical Society Archives.

25. "Shiny New Coat of Aluminum Paint for Bear Mountain Bridge," *Peekskill (NY) Evening Star*, July 11, 1949.

26. Smith to Harriman, "Bear Mountain Hudson River Bridge Company Monthly Progress Report," Nov. 1, 1924, Harriman Collection, Orange County Historical Society Archives.

27. Smith to Harriman, "Bear Mountain Hudson River Bridge Company Monthly Progress Report," Nov. 1, 1924, Harriman Collection, Orange County Historical Society Archives.

28. Harriman to Tench Construction, Jan. 7, 1925, Harriman Collection, Orange County Historical Society Archives.

29. "Frederick Tench, a Steel Engineer," *New York Times*, Oct. 28, 1944.

30. "Mr. Pugsley Lauds Builder of Bridge," *Yonkers (NY) Herald*, Dec. 11, 1924.

31. "Bear Mountain Bridge," *New York Times*, Nov. 28, 1924.

20. The Bond

1. E. R. Harriman to Charles H. Sabin (and others), Aug. 2, 1928; George W. Perkins Jr. to Harriman, Aug. 14, 1929, Harriman Collection, Orange County Historical Society Archives.

2. Harriman to several investors and responses, 1928–29; Perkins to Harriman, Aug. 14, 1929, Harriman Collection, Orange County Historical Society Archives.

3. Harriman to investors, 1928–29, Harriman Collection, Orange County Historical Society Archives.

4. Perkins to Harriman (attachment to correspondence listing of investors participating in Harriman's plan), Aug. 14, 1929; Perkins to Harriman, Aug. 14, 1929, Harriman Collection, Orange County Historical Society Archives.

5. Perkins to Harriman, Aug. 14, 1929, Harriman Collection, Orange County Historical Society Archives.

6. Perkins to Harriman, Aug. 14, 1929, Harriman Collection, Orange County Historical Society Archives.

7. "Cross Hudson Span in Autos This Week," *New York Times*, Nov. 23, 1924.

8. Henry J. Stanton, "View: Bear Mountain Bridge's Secret History," *LOHUD*, Nov. 27, 2014, https://www.lohud.com/story/opinion/contributor/2014/11/27/bear-mountain-bridge-secret-history/19394987/.

9. Wolf, *Crossing the Hudson*, 141, 178.

10. "More Typhoid Is Traced to Brook," *Passaic (NJ) Daily Herald*, July 25, 1924.

11. W. H. Averell to Harriman, May 15, 1925, Harriman Collection, Orange County Historical Society Archives.

12. Indenture document, 1936, between Bear Mountain Hudson River Bridge Company and holders of said company's income bonds (to reduce the interest rate), Harriman Collection, Orange County Historical Society Archives.

21. Remembering

1. Ople Beall (on behalf of William A. Welch) to L. S. Osborne, June 20, 1928, Palisades Interstate Park Commission (hereafter cited as PIPC) Historical Archives.

2. Osborne to Welch, Sept. 24, 1928, PIPC Historical Archives.

3. "Montclair Women Participate Memorial Celebrating Palisades," *Montclair (NJ) Times*, May 4, 1929.

4. "Program for the Dedication of Federation Memorial Park—Alpine, N.J., Tuesday, April 30th, 1929," New Jersey State Federation of Women's Clubs (hereafter cited as NJSFWC) Historical Archives.

5. "Biography of Cecilia Gaines Holland," written by a club member, NJSFWC Historical Archives.

6. Binnewies, *Palisades*, 238, 241.

7. Baird [author assumed], "Bear Mountain Hudson River Bridge."

8. Baird [author assumed], "Bear Mountain Hudson River Bridge."

9. Baird [author assumed], "Bear Mountain Hudson River Bridge."

10. Baird [author assumed], "Bear Mountain Hudson River Bridge."

11. Baird [author assumed], "Bear Mountain Hudson River Bridge."

12. E. Roland Harriman to Howard C. Baird, Nov. 29, 1924, Harriman Collection, Orange County Historical Society Archives.

13. Baird to Harriman, Dec. 1, 1924, Harriman Collection, Orange County Historical Society Archives.

14. Records of the Institution of Civil Engineers, membership documents, 1929, https://www.ancestry.com/discovery.

15. Burrows and Wallace, *Gotham*, 713.

16. Burrows and Wallace, *Gotham*, 713.

17. Felicia R. Lee, "121 Years of Men Only Ends at Club," *New York Times*, July 28, 1989.

18. "Howard Carter Baird, Biographical Archive: Earliest Members of the Century Association," Century Association Archives Foundation.

19. Sherman Baldwin, treasurer at the Century Association, to James B. Jacob, May 23, 1957, Century Association Archives Foundation.

20. "Howard C. Baird," obituary, *Louisville (KY) Courier-Journal*, Dec. 13, 1957.

22. Update—1927 through the Present

1. "Morgan Named Park Manager," *Bergen (NJ) Evening Record*, Feb. 2, 1940.

2. "George Perkins, Ex-U.S. Aide, Dies," *New York Times*, Jan. 12, 1960.

3. "George W. Perkins," in *Who's News and Why*.

4. "George W. Perkins—Merck Mourns."

5. "George W. Perkins—Merck Mourns."

6. "Parkway Opens Final Section," *Bergen (NJ) Evening Record*, Aug. 29, 1958.

7. Binnewies, *Palisades*, 71.

8. "Perkins Dead; U.S. Diplomat," *Bergen (NJ) Record*, Jan. 12, 1960.

9. "Tribute to George W. Perkins," *Pottstown (PA) Hill News*, Jan. 22, 1960.

10. Binnewies, *Palisades*, 51–54.

11. Abramson, *Spanning the Century*, 237.

12. Veronica Campaniello, "Mary Harriman Rumsey: A Woman Ahead of Her Time," *Port Washington (NY) Weekend*, Sept. 2–8, 2015.

13. Abramson, *Spanning the Century*, 384, 588, 697.

14. Harriman, *I Reminisce*.

15. "E. Roland Harriman Is Dead at 82—Financier and Trotting Sponsor," *New York Times*, Feb. 2, 1978.

16. Harriman, *I Reminisce*.

Epilogue

1. Bear Mountain Hudson River Bridge Company Charter, issued by the state of New York, Mar. 1922, Palisades Interstate Park Commission Historical Archives.

2. Indenture transferring title for the Bear Mountain Bridge from Bear Mountain Hudson River Bridge Company to the people of the state of New York, Sept. 25, 1940, New York State Library Archives.

3. Cameron, *Spanning the Hudson*.

4. Cameron, *Spanning the Hudson*.

5. "Bear Mountain Bridge Centennial," Bridge Authority of New York, 2024, https://www.nysba.ny.gov/bmb100.

6. "Bear Mountain Bridge to Celebrate Centennial with Motorcade and Documentary Film," *Peekskill (NY) Herald*, Nov. 21, 2024.

Bibliography

Newspapers

Bergen (NJ) Evening Record, 1940–58
Bergen (NJ) Record, 1960
Bernardsville (NJ) News, 1922
Brooklyn Citizen, 1907
Brooklyn Daily Eagle, 1897–1924
Elmira (NY) Star-Gazette, 1930
Gloucester County (NJ) Democrat, 1900
Harper's Weekly, 1890
Jersey City (NJ) News, 1894–97
Louisville Courier-Journal, 1957
Middletown (NY) Daily Herald, 1924
Montclair (NJ) Times, 1929
Newark Daily Advertiser, 1901
New York Daily Tribune, 1894–1920
New York Evening Post, 1920
New York Globe, 1917
New York Herald, 1920–22
New York News, 1899
New York Sun, 1897–1919
New York Times, 1889–1978
New York Tribune, 1894–1920
New York World, 1897
Niagara Falls (NY) Review, 1963
Passaic (NJ) Daily Record, 1924
Peekskill (NY) Evening Star, 1924
Peekskill (NY) Highland Democrat, 1923–24
Perth Amboy (NJ) Evening News, 1907–23
Philadelphia Times, 1899
Port Washington (NY) Weekend, 2015
Pottstown (PA) Hill News, 1960
Poughkeepsie (NY) Eagle-News, 1922
Scarsdale (NY) Inquirer, 1923
Washington (DC) Evening Star, 1917
Yonkers (NY) Herald Statesman, 1922
Yonkers (NY) Herald, 1924

Archives

American Society of Civil Engineering 46, Part 1. Jan. 1920. https://archive.org/details/proceedings46amer/page/n5/mode/2up. Accessed Nov. 2024.

Century Association Archives Foundation, New York.

Harriman, E. Roland. Personal Papers. Rare Books & Manuscripts Library, Butler Library, Columbia Univ., New York.

Harriman Collection. Orange County Historical Society Archives, Arden, NY.
Historical Archives of the Women's Club of Englewood, Englewood, NJ.
"The History of Ironworkers Local 361." Historical Archives of Ironworkers Local 361, New York.
New Jersey State Federation of Women's Clubs Historical Archives. New Brunswick, NJ.
New York State Library Archives. Albany, NY.
Palisades Interstate Park Commission Historical Archives. PIPC of New York and New Jersey, Iona Island, NY.
Pennsylvania, US, National Guard Veterans Card Files, 1867–1921. https://www.ancestry.com/discovery. Accessed Nov. 2024.
Perkins, George W., Jr. Personal Papers. Rare Books & Manuscripts Library, Butler Library, Columbia Univ., New York.
Perkins, George W., Sr. Personal Papers. Rare Books & Manuscripts Library, Butler Library, Columbia Univ., New York.
Phoenix Bridge Company. Published Collections. Hagley Library, Wilmington, DE. Accession 916.
Records of the Institution of Civil Engineers, London. https://www.ancestry.com/discovery. Accessed Nov. 2024.

Other Sources

Abbott, Lawrence. "Comments on the Bear Mountain Bridge." *Outlook*, July 18, 1923.
Abramson, Rudy. *Spanning the Century: The Life of W. Averell Harriman, 1891–1986*. New York: W. Morrow, 1992.
"Achievements of George W. Perkins." *New York National Civic Federation Review* (July 10, 1920).
An American author. *American Husbandry: An Account of the Soil, Climate, Production, and Agriculture of the British Colonies in North America and the West Indies*. London: J. Bew, 1775, courtesy of the Hudson River Maritime Museum. https://www.hrmm.org.
Baird, Howard Carter [author assumed]. "Bear Mountain Hudson River Bridge." Engineering diss. and assessment, ca. 1925. Palisades Interstate Park Commission Historical Archives.
"The Bear Mountain Bridge." *Outlook* 138, no. 15 (1924): 582.
"Bear Mountain Bridge—75th Anniversary." New York Bridge Authority, 1999.

"Bear Mountain Inn." *American Architect* 108 (Nov. 10, 1915).
Binnewies, Robert O. *Palisades: 100,000 Acres in 100 Years*. New York: Fordham Univ. Press and Palisades Interstate Park Commission, 2001.
Birmingham, Stephen. *Our Crowd: The Great Jewish Families of New York*. Syracuse, NY: Syracuse Univ. Press, 1996.
"Bridge Design and Good Taste." *Engineering News-Record* (July 2, 1923).
Burrows, Edwin G., and Mike Wallace. *Gotham: A History of New York City to 1898*. New York: Oxford Univ. Press, 1999.
Cameron, Leslie Paige. "Spanning the Hudson: A History of the New York Bridge Authority." New York Bridge Authority, n.d.
Carr, William H. "Signs along the Trail." *Dept. of Education, American Museum of Natural History*, New School Service ser., no. 2 (1927).
Clark, Donald F. "The Bridge That Never Was: Precursor of the Bear Mountain Hudson River Bridge." *Journal of the Orange County Historical Society*, no. 7 (1977–78).
Commissioners of the Palisades Interstate Park. *The Palisades Interstate Park, 1900–1929: A History of Its Origin and Development*. New York: Palisades Interstate Park Commission, 1929.
Construction of Parallel Wire Cables for Suspension Bridges. Trenton, NJ: John A. Roebling's Sons, 1925.
"Construction of the Lincoln Memorial." National Park Service, last updated May 18, 2021. https://www.nps.gov/linc/learn/historyculture/lincoln-memorial-construction.htm.
"Design of Bear Mountain Bridge Attacked as Ugly." *Engineering News-Record* (July 26, 1923).
Dunwell, Frances F. *The Hudson River Highlands*. New York: Columbia Univ. Press, 1991.
"Edward F. Terry Obituary." *Engineering News-Record* 92 (1924).
Evans, Tony Tekaroniake. "How Mohawk 'Skywalkers' Helped Build New York City's Tallest Skyscrapers." History.com, May 13, 2021. https://www.history.com/news/mohawk-skywalkers-ironworkers-new-york-skyscrapers.
Figliomeni, Michelle P. *E. H. Harriman at Arden Farms*. Arden, NY: Orange County Historical Society, 1997.
Garraty, John A. *Right Hand Man: The Life of George W. Perkins*. New York: Harper & Bros., 1957.
"George W. Perkins." In *Who's News and Why*, vol. 1, no. 4. New York: H. W. Wilson, 1950.

"George W. Perkins—Merck Mourns." *Merck Company Newsletter* (Jan. 1960).
Gibson, Chelsea. *Biographical Sketch of Adaline Wheelock Sterling*. Binghamton, NY: Binghamton Univ., accessed through the Women's Club of Englewood Historical Archives.
Griggs, Frank, Jr. "Great Achievements: Othmar H. Ammann." *Structure Magazine* (Apr. 2013): 42–44.
Harper Bennett, William. *Catholic Footsteps in Old New York: A Chronicle of Catholicity in the City of New York from 1524 to 1808*. New York: Schwartz, Kirwin, and Fauss, 1909.
Harriman, E. Roland. *I Reminisce*. Garden City, NY: Doubleday, 1975.
"History of Harriman State Park of Idaho." Idaho Dept. of Parks and Recreation, updated Summer 2024. https://parksandrecreation.idaho.gov/parks/harriman/history/.
"Hudson River Bridge at Bear Mountain: A Solution of the Motor Traffic Problem across the Upper Hudson River." *Scientific American*, May 1923.
Jeffers, H. Paul. *The Bully Pulpit: A Teddy Roosevelt Book of Quotations*. Dallas: Taylor, 2002.
Josephson, Matthew. *The Robber Barons: The Great American Capitalists, 1861–1901*. New York: Harcourt Brace Jovanovich, 1962.
Kennan, George. *Railroad Tycoon: A Biography of E. H. Harriman*. 1922. Reprint, Big Byte Books, 2014. Kindle.
Klein, Maury. *The Life and Legend of E. H. Harriman*. Chapel Hill: Univ. of North Carolina Press, 2000.
Kyvig, David E. *Daily Life in the United States, 1920–1940: How Americans Lived through the Roaring Twenties and the Great Depression*. Chicago: Ivan R. Dee, 2003.
Leonard, John, ed. *Woman's Who's Who in America*. New York: American Commonwealth, 1914.
Lewis, Tom. *The Hudson: A History*. New Haven, CT: Yale Univ. Press, 2005.
Mack, Arthur Carlyle. *The Palisades of the Hudson: Their Formation, Tradition, Romance, Historical Associations, Natural Wonders and Preservation*. Edgewater, NJ: Palisades Press, 1909.
MacKaye, Benton. "An Appalachian Trail, a Project in Regional Planning." *Places Journal* (Apr. 2019). Reprint of the original article written by MacKaye in the 1920s with an introduction by Garrett Dash Nelson. https://placesjournal.org/article/an-appalachian-trail-a-project-in-regional-planning/.

Maddox, Kenneth W. "The Lure of the Country." In *Westchester: An American Suburb*, edited by Roger Panetta, 103–36. New York: Fordham Univ. Press, Hudson River Museum, 2006.

Manoff, Arnold, and Chris Thorsten. "Random Interviews of Union Ironworkers Regarding Job Issues and Coworkers." New York, 1938. Manuscript/Mixed Material. https://www.loc.gov/item/wpalh001462/.

McCully, Betsy. *City at the Water's Edge: A Natural History of New York*. New Brunswick, NJ: Rivergate, an imprint of Rutgers Univ. Press, 2007.

McSherry, Patrick. "The History of the 3rd U.S. Volunteer Engineers." Spanish American War Centennial website. https://www.spanamwar.com/3rdengineers.htm. Accessed Nov. 2024.

"Memoir of Henry Wilson Hodge." *Proceedings of the American Society of Civil Engineers* 46, no. 1 (1920): 703–4.

Middleton Dwyer, Michael, ed. *Great Houses of the Hudson River*. Boston: Bulfinch, 2001.

Muir, John. *Tribute to Edward Henry Harriman*. Garden City, NY: Doubleday, Page, 1912.

Nash, Roderick Frazier. *Wilderness and the American Mind*. New Haven, CT: Yale Univ. Press, 1967.

"Othmar Hermann Ammann." American Society of Civil Engineers. https://www.asce.org/about-civil-engineering/history-and-heritage/notable-civil-engineers/othmar-hermann-ammann. Accessed Jan. 31, 2025.

The Palisades Interstate Park, 1900–1929: A History of Its Origin and Development. Commissioners of the Palisades Interstate Park, 1929.

Palisades Interstate Park Commission. *60 Years of Park Cooperation*. Bear Mountain, NY: Palisades Interstate Park Commission, 1960.

Panetta, Robert. *Westchester: The American Suburb*. New York: Fordham Univ. Press and Hudson River Museum, 2006.

Partridge, Dr. E. L. "A National Park on the Hudson." *Outlook* 87 (Nov. 9, 1907).

Petroski, Henry. *Engineers of Dreams: Great Bridge Builders and the Spanning of America*. New York: Vintage Books, 1995.

Reier, Sharon. *The Bridges of New York*. New York: Quadrant Press, 1977.

Schley, Ben. "A Century of Fish Conservation." National Conservation Training Center, US Fish and Wildlife Service.

Schuyler, David. *Sanctified Landscape: Writers, Artists, and the Hudson River Valley, 1820–1909*. Ithaca, NY: Cornell Univ. Press, 2012.

Smith, Wilson Fitch. "Bridging the Hudson River at Bear Mountain." *Engineering News-Record* (May 10, 1923).

Torrey, Raymond H., Frank Place Jr., and Robert L. Dickinson. *New York Walk Book*. New York: Dodd, Mead, 1934.

"Towers of First Suspension Bridge over Hudson River Completed." *Engineering News-Record* (Apr. 3, 1924).

Trudeau, Edward Livingston. *To Comfort Always*. Big Byte Books, 2014. Kindle.

"Tugboats: Workhorses of the Hudson River." Hudson River Maritime Museum/New York Heritage Digital Collections. https://nyheritage.org/collections/tugboats-workhorses-hudson.

US Fish and Wildlife Service. "History of the U.S. Fish and Wildlife Service." https://www.fws.gov/history-of-fws. Accessed Dec. 17, 2024.

Wagner, Erica. *Chief Engineer: Washington Roebling, the Man Who Built the Brooklyn Bridge*. New York: Bloomsbury, 2017.

Winpenny, Thomas R. *Without Fitting, Filing, or Chipping: Illustrated History of the Phoenix Bridge Company*. Easton, PA: Canal History and Technology Press, 1996.

Wolf, Donald E. *Crossing the Hudson: Historic Bridges and Tunnels of the River*. New Brunswick, NJ: Rivergate Books, an imprint of Rutgers Univ. Press, 2010.

Index

Italic page number denotes illustration.

Abbott, Lawrence, 172, 173
abutments, 177, 179, 180, 182, 183, 186, 187, 217
Adams, Charles Frances, 149
Adams, John, 106
Adams, John Quincy, 106
air compressors, 16, 175, 185
Albert, Fleming P., 34
Algonquin, 20. *See also* Iroquois
American Forestry Association, 29
American Husbandry, 21
American Revolutionary War, 21
American Scenic and Historic Preservation Society, 34
Amman, Otmar H., 173
anchorages, 141, 176, 179–82, 183, 186, 187, 189, 212, 230, 231, 248; tunnels, 175, 177, 179, 180, 214–15
anchor bars, 174, 182
Appalachian Mountain Club, 29
Appalachian National Scenic Trail, 113
approach spans, 180, 187–88, 190, 191, 214
Arden Farms Dairy Company, 47
Arden Farms General Stores, 47
Arnold, Clinton S., 33
artistic movements, 4
As You Like It (Shakespeare), 46

automobiles, 124–25; purchase prices, 125; traffic, 125–26
Averell, William, 45

Baird, Howard Carter, 5, 14, 15, 62, 134–38, 143, 174, 179, 188, 214, 230–35; Boller & Hodge, 137, 234; Century Club, 234–35; death, 235; design documents, 134; dissertation, 137, 173; family background, 135; McClintic Marshall plants, 175; as member of engineering organizations, 234; military service, 136; organizing civil engineers as a professional society, 233; Phoenix Bridge Company, 134–37, 232, 234; Roland's letter to, 232–33; United Kingdom, 234
Baird, James P., 135
Baird, Martha Howard, 135
Ball, Evelina. *See* Perkins, Evelyn
Bannard, Otto, 169
Barrett, Nathan, 77
basalt, 57
base castings, 176, 179–81
Beall, Camille, 107
Beall, E. W., 107

284   Index

beams, 153; floor, 190, 214
Bear Mountain Bridge (BMB), 2–3, 95, 224; campaign for, 95; construction (*see* construction (BMB)); declining traffic and revenue, 225–26; documentary film, 249; history, 2; inscription, 10; location, 2; news articles/reports, 10–11, 12, 15–16, 172; opening, 8–17, 218–19; parallel wire-cable construction, 185–86; toll plaza, 247
Bear Mountain Bridge: The First 100 Years, 249
Bear Mountain Hudson River (BMHR) Bridge Company, 9, 12, 16, 138, 217, 226, 245; annual stockholders meeting, 169–70; board of directors, 145; bonds (*see* bonds); Certificate of Incorporation, 144; charter, 141–45; mortgage and deed of trust, 169; plans and specifications, 160–61; sinking-fund provision, 167, 169; Terry & Tench Co., Inc., and, 167
Bear Mountain Inn, 12, 107–8, 218, 226, 249
Bear Mountain State Park, 95, 105–6, 108, 247
Bear Mountain Trail, 112–13
Beers, William H., 55
Belknap, John M., 232
Benjamin Franklin Bridge, Philadelphia, 8, 224
Bergen County Historical Society, 89
Bernardsville (NJ) News, 141
Berwind, E. J., 165–66
Bierstadt, Albert, 21–22
Bilovick, Steve, 215
blasts/blasting, 32–33
boats, 110

Boller, Alfred P., 137
Boller & Hodge, 137, 234
Bond, William E., 30–31
bonding company, 162, 178
bonds, 167, 169; debenture, 143, 163–66, 226; first-mortgage, 143, 163, 169; income, 163, 166, 169, 221–23; performance, 16, 161, 162, 208, 209, 222; surety, 162
Book of Englewood, The (Sterling), 91
Boys Club of Tompkins Square, 44–45
Bradley, Stephen Rowe, 103
British colonies, 20–21
Brooklyn Bridge, 84–85, 224–25
Brooklyn Daily Eagle, 71–72, 139
Brown, Edward F., 113
Bryant, William Cullen, 234
Buck, Leffert, 173, 186

cables, 1, 2, 140, 177, 189–91, 207–8, 211–14, 217, 225, 230–32; footbridge, 187, 189, 213; hanger, 179; parallel wire construction, 185–86, 207; suspender, 168, 174, 182, 190, 231
Campbell, Helen, 181
Campfire Girls, 109
camping, 96, 109–10
Carpathia, 23
Carpenter Brothers quarry, 32, 78
Carr, William H., 113
Casey, Edward P., 235
catenary, 2, 174
Catskill Aqueduct, 93–94
Catskill Mountains, 93–94
Century Club/Association, 234–35
children, 109
Church, Frederic, 21–22
Clark, George C., 44

Index   285

Clermont, 23, 93
Coe, George S., 30
cofferdam, 171, 176, 179, 181
Cole, Thomas, 21–22
communication, 28
Constitution Island, 102
construction (BMB), 4, 11, 171–91, 192–206; anchorage tunnels, 177, 214–15; base castings, 176, 179–81; cofferdam, 171, 176, 179, 181; concrete roadway, 214; contractor billing, 178; delays, 171–72, 175–78; design, 172–73; excavation, 16, 171, 175–77, 179, 180–83, 185–87, 211, 213, 215; fabrication, 174–76, 178–79, 183, 185–86, 207, 211; fatalities, 215; flat-wire stay, 207–8; McClintic Marshall Co., 171, 174–76, 178, 187, 189, 191, 211, 213; PIPC and, 174, 177; power plant, 175; precast concrete slabs, 214; progress reports, 174–76; public attention, 172; schedule, 172, 177, 208
Coppage, Riley, 190
Couch, Carolyn, 177
Couch, Clifford, 177
Couch, Franklin, 177
Couch, Mary, 177
Couch swamp, 188, 190–91
creeper traveler, 184–85
Cropsey, Jasper Francis, 21–22
Croton Aqueduct, 24
Cutler, Otis H., 121

Davidson, M. O., 27
Dawson, Henry H., 88
Dawson, Ida W., 123
D. C. Hays and Co., 43–44

debenture bonds, 143, 163–66, 226
deck/floor systems, 183, 189
deck suspenders, 174
Delaware, 20. *See also* Lenape people
Demarest, S. Elizabeth, 122, 228
DeMastini, Maries, 215
Denver Pacific, 149
De Ronde, Abram, 77, 88
derricks, 182, 184, 185
Dickinson, Robert L., 112
Dobson, Meade C., 111
Dodd, Samuel B., 31
Donohue, Charles J., 9
Douglas, James, 103

Economic Cooperation Administration (ECA), 241
Ed (Perkins's colleague), 159–61
Ellis, George, 142–46, 159, 162, 164, 166–67
Ellison, William O., 32
Emery, Henry G., 123
Engineering News-Record, 167–68, 172, 184, 210
Englewood Women's Club, 66, 68, 82, 89, 122, 229
Erie Canal, 22
Erie Railroad, 26
eyebars, 179, 180, 181, 212

Farnam, Tom, 168–69
first-mortgage bonds, 143, 163, 169
fishing villages, 23
floor beams, 190, 214
Focht, Jack, 114
footbridge cables, 187, 189, 213
footbridges, 187, 189–90, 191, 214, 231

Fries, Harold, 131
Fulton, Robert, 23, 93

gag press, 135
Gaines, Cecilia. *See* Holland, Cecilia Gaines
Gaines, Henry, 74
Gaines, Jane A., 74
G. Amsinck and Company, Inc., 159
Garrison, Wendell P., 31
General Motors Acceptance Corporation, 125
George Washington Bridge, 225
Globe Insurance, 162, 186, 209
Golden Age of Bridges, 224–25
Gourley, William B., 58
Grant, Ulysses S., 26
Great Depression, 226
Great War. *See* World War I
Green, Andrew H., 69, 70, 72
Griffen, John, 135

Half Moon, 19, 93
Hall, John, 159–61
hanger cables, 179
Harper's Bazaar, 90
Harper's Weekly, 27
Harriman, Cornelia Nielson, 42
Harriman, Edward Henry, 5, 9, 35, 41–51, 86, 131, 243; charitable work, 44–45; dairy business, 47; death, 101, 130; descendants, 245; education/school, 42–43; family background, 42; financial/stock market, 43–44; health issues, 100; horses and, 48; Kahn on, 49; Klein's biography of, 101; marriage, 45; Morgan and, 49–51, 101; Muir on, 129–30; Northern Securities Holding Company, 50–51; Partridge and, 99; railroads, 45–47, 124, 129, 131; river crossings, 124; Roosevelt and, 51; Schiff and, 49, 50, 86, 150; social circles, 44; Tower Hill mansion, 86–87; wealth, 131
Harriman, E. Roland, 9, 12, 100, 124, 128–32, 218, 244–45; bridge investors and, 221–24; business ventures, 158; death, 245; father and, 129–30; financial program, 95, 162; health issues, 132; *I Reminisce*, 129, 130, 165; marriage, 131; philanthropic affiliations, 245; privilege, 129; railroads, 157–58; Santa Barbara, 132; schooling, 130; US Army, 116, 132. *See also* construction (BMB)
Harriman, Gladys Fries, 131, 132, 244, 245
Harriman, Mary Averell, 9, 45, 100, 101–4, 128, 157–58, 243; shipbuilding, 132
Harriman, Oliver, 44
Harriman, Orlando, 42, 43
Harriman, William Averell, 103, 128, 244
Harriman State Park of Idaho, 130
Harriman State Park of New York, 130
Haverstraw, 20
Hayes, Will, 120
Hazard, Rowland, 181
H. D. Ward and Company, 25
Herpt, John, 216
Hewitt, Abram S., 77
hiking trails. *See* trails/trail system
Hodge, Henry W., 137, 230
Holland, Cecilia Gaines, 73–75, 228–29; League for the Preservation of the Palisades, 85–86; "The Saving of the Palisades," 73

Holland, John A., 75, 229
Hopkins, Franklin W., 77
Hopper, John, 58
Hornbostel, Henry, 173
Horne, H. Field, 9
Howard, Charles E., 99
Howell, William Thompson, 29
Hubbard, Louis V., 228
Hudson, Henry, 19, 93
Hudson-Fulton Celebration, 92–93
Hudson Highlands, 18
Hudson River, 99; *American Husbandry* on, 21; as "Devil's Horse Race," 99; magnificence, 22; as means of travel and commerce, 21; railroad bridges, 124; waterfront, 22
Hudson River Railroad, 25–26
Hudson River School, 4, 22, 234
Hudson Valley: economic boom, 23
Hughes, Charles Evans, 100, 101, 103

income bonds, 163, 166, 169, 221–23, 274n12
Indigenous people, 19–20
Industrial Revolution, 23
Inglis, James, Jr., 58
Inness, George, 21–22
iron ore, 42
ironworkers, 152–55; accidents, 216–17; death/fatalities, 184; movement, 153–54; nonunion, 153, 154; standard wage, 154; strike, 154–55; union, 153, 154
Iroquois, 20. *See also* Algonquin

Jacob, James Baird, 235
Jarrett, Edwin S., 235
Jersey City News, 30, 33

Jersey City Women's Club, 74
Jervis, John Bloomfield, 24–25
John A. Roebling's Sons Company, 168, 174, 179–82, 185, 187, 191, 211, 213, 231, 232
Jolliffe, Ruby M., 105, 114
Journal of the American Institute of Architects, 112

Kansas Pacific, 149
Kellogg, Clarke, and Company, 135. *See also* Phoenix Bridge Company
Kennan, George, 43
Kennedy, John F., 244
kinship, 20
Kittredge, George W., 186
Klein, Maury, 101
Knickerbocker, 22
Knickerbocker Group, 4, 22
Kuhn, Loeb, and Co., 150

labor unions, 152–53
Lake Ontario Southern Railroad, 45–46
Lamb, Frederick, 84
League for the Preservation of the Palisades, 79–82, 84, 85, 88, 89, 122
League of Walkers, 111
Leavitt, Charles Wellford, 98–99, 105
Lenape people, 19–20; autonomous groups, 20; kinship, 2
Leopold, Victor, 232
Lincoln Memorial, 157
Lindenthal, Gustav, 173
Linn, W. A., 77
Loomis, Sarah Sophia Dana, 66–70, 89, 228–29
Lovett, Bob, 131

Lovett, Robert, 158
Lusitania, 23

MacKaye, Benton, 112–13
Manhattan, 19
Manhattan Bridge, 173, 180, 217
Marshall Plan, 213
Mastick, Seabury C., 141
Mather, Stephen Tyng, 110
Mathis Yacht Building Company, 110
Mayflower, 106
McCall, John A., 55, 56, 76
McClintic Marshall Co., 171, 174–76, 178, 187, 189, 191, 211, 213
McKay, William J., 104
McKinley, William, 51
McKinnon, Joseph, 216
Mercator, Gerardus, 18, 19
Merck, Caroline (Linn). *See* Perkins, Caroline (Linn) Merck
Merck, Friedrich Jacob, 120
Merck, George, 120–21
Merck, George W., 120
Merck and Company, 120–21
Miles, E. B., 89
Miller, Nathan, 121
mines, 42
Mitchel, John Purroy, 116
Model T, 125
Modjeski, Ralph, 180
Mohawk nation, 183–84
Mohawks, 20
Morgan, J. P., 49–51, 52, 67, 79, 82, 83, 86, 92, 101, 102, 104, 154
Morgan, Kenneth, 240
Morgan's Raiders, 106
Morton, Levi P., 58
Morton, L. L., 232
motorcars. *See* automobiles

Muir, John, 129–30
Munsee, 20

Nation, 31
National Audubon Society, 29
"National Park on the Hudson, A" (Partridge), 99
National Park Service, 110, 248
National Park System, 113
National Recovery Administration, 244
National Register of Historic Places, 248
NATO (North Atlantic Treaty Organization), 241, 242
natural resources: conservation, 28–34; exploitation, 28
New England Railroad, 26
New Jersey Bulletin, 73
New Jersey State Federation of Women's Clubs (NJSFWC), 3, 74; Committee on Forestry and for the Protection of the Palisades, 89; Memorial Committee, 227; *New Jersey Bulletin*, 73
Newsweek, 158, 244
New York Association for Improving the Condition of the Poor, 109
New York Central Railroad, 46, 150, 152, 174–75, 188, 208, 209
New York Evening Post, 111, 112
New York Life Insurance Company (NYL), 52–57, 76, 79, 82
New York State Bridge Authority (NYSBA), 4, 247–48
New York Stock Exchange, 44
New York Times, 10, 58, 85, 104, 172, 220
New York Tribune, 32, 87
New York Walk Book, 112
Nobel, Alfred, 98
Northern Securities Holding Company, 50–51

O'Brien, Morgan J., 15
Odell, Benjamin B., 15, 104, 216
Odell-Foss Quarry Company, 33
Ogdensburg and Lake Champlain Railroad Company, 45
Old Volumes Book Club, 74
Olmsted, Frederick Law, 34
Orange County Historical Society, 4
Osborne, Henry Fairfield, 110
Osborne, Lydia S., 123, 227
Outlook, 17, 99, 172
overindustrialization, 6; protection against, 29

Palisades, 18; destruction of, 31, 35
Palisades (steamship), 110
Palisades Interstate Park, 31; concerns, 96; dedication, 93; as recreational destination, 96
Palisades Interstate Park Commission (PIPC), 3–4, 110; Carpenter quarry purchase, 78; commissioners, 77–78, 92, 122, 240; donations, 79, 82, 83, 92, 101–4; goals and aspirations, 98; headquarters, 77; Hudson-Fulton Celebration, 92–93; innovative programs, 226; jurisdiction, 87; Manhattan Trap Rock properties, 103; patrons, 92; political resistance, 97; professionalism, 92; property holdings, 113; quelling trespassers, 83; regulations, 96; Storm King Highway, 94; Township of Alpine tax bills, 96–97; trail system, 111–12; treasury, 77
Palisades Interstate Parkway, 97
Palisades Mountain House, 23
Palisades Railroad, 30
parallel wire-cable construction, 185–86, 207

parapet, 215
parkways, 97
Parrott, Edward, 41
Parrott, Peter, 42
Parrott, Robert Parker, 42
Partridge, Edward Lasell, 99, 104
Patriot, 90
Peekskill (NY) Evening Star, 189
Pennsylvania Railroad, 46
performance bonds, 16, 161, 162, 208, 209, 222
Perkins, Caroline (Linn) Merck, 120–21
Perkins, Dorothy, 55, 57, 104
Perkins, Evelyn, 54–56, 76, 104, 117
Perkins, George W., 53
Perkins, George W., Jr., 14, 56, 114, 115–21, 221, 240–43; college years, 115–16; federal government jobs, 115; PIPC, 115, 121, 159; Tench and, 159–60, 209–10; US Army, 115, 116–17; YMCA, 116, 118–19
Perkins, George W., Sr., 3, 37, 51, 115; childhood, 53; death, 119–20; family background and legacy, 53; health issues, 117–19; Katherine (daughter-in-law) and, 117, 118; leadership, 54–55; Leavitt and, 98–99; marriage, 54; memorials, 120; Morgan and, 79, 82, 83; National City Bank, 78–79; NYL, 52–57, 76, 79, 82; PIPC, 52, 76–77, 83, 84, 86, 118; Red Triangle program, 118; Republican Party, 116; Riverdale-on-Hudson, 52, 55, 57; Roosevelt and, 76; United War Work Council, 117; Vermilye and, 79–82, 84; Welch and, 105, 107–8, 114, 118, 120; YMCA, 117, 118
Perkins, Katherine Trowbridge, 116, 117
Perkins, Sarah, 53
Perkins, Willie, 53

Perth Amboy (NJ) Evening News, 33
Phelps, William Walter, 30
Phipps, Henry, 104
Phoenix Bridge Company, 134–37, 232, 234
Phoenix Column, 135
Phoenix Steel Corporation, 135
piers, 140, 177, 179, 180–81
Place, Frank, Jr., 112
plazas, 179, 180, 186, 187
Poughkeepsie Eagle News, 139
prison plan (New York State), 99–100, 102, 104
Proctor, Olin S., 155
progress payments, 180
Pugsley, Cornelius A., 11–12, 219

railroads/railways, 23–26; charter, 24–25; growth, 29, 123; Edward Henry Harriman, 45–47, 124, 129, 131; negative impact, 29; passengers, 126. *See also specific railroads*
Rare Book & Manuscript Department of Butler Library, 4
Raritan tribe, 20
Red Triangle program (YMCA), 118
Reeves, David, 135
Reeves, Samuel J., 135
retainage, 164, 178
retaining walls, 16, 119, 180, 183, 211, 215
Reynders, John V. W., 13–14, 186–88, 208
Riney, Michael, 148
Riverdale-on-Hudson, 52, 55, 57
riveting, 183–84, 187
Roberts, Emma Harriet. *See* Tench, Emma Harriet Roberts
Roberts, John, 150

Roberts, Sarah J. McKernan, 150
Robinson, Holden D., 14, 232
Rockefeller, John D., 51, 92, 102
Rockefeller, John D., Jr., 242
Rockefeller, Nelson, 244
Rockland Lake Trap Rock Company. *See* Odell-Foss Quarry Company
Roebling, Emily Warren, 84–85
Roebling, Washington A., 84–85, 232
Roosevelt, Franklin D., 247
Roosevelt, Theodore, 34, 51; conservation movement, 72; as governor, 72; Perkins and, 76; PIPC and, 76
Rossi, Irving, 16
Rumsey, Mary Harriman, 243–44
Russell Sage Foundation, 102

Sage, Margaret Slocum, 102
Sage, Russell, 102
Sauzade, Katharine, 85
Schiff, Jacob, 49, 50, 86, 150
Scientific American, 125, 140
Second Industrial Revolution, 28
Serrell, Edward W., 26–27
Sherman Anti-trust Act, 51
Sherwin, H. H., 137–38
shop drawings, 134, 174, 234
Sierra Club, 29
Sloane, Sam, 25
Smith, Alfred E., 12
Smith, C. Ernest, 141
Smith, George T., 121
Smith, Kate, 108
Smith, Wilson Fitch, 5, 9, 10, 167–68, 171, 174–75, 217–18
Smith, W. T., 160–61
Smith-Mastick bill, 140, 141, 143–44
Smock, John, 31
Sodus Bay and Southern Railroad, 46

Index   291

SS *Obidense*, 107
staging areas, 188
Stanton, Elizabeth Cady, 90
Stauffer, D. McNeeley, 77
steamboats, 23, 24; transportation monopoly, 24
steam locomotives, 23
steam power, 23
steam shovels, 175
Sterling, Adaline Wheelock, 66, 68, 69, 89–91, 228–29; Englewood Board of Education, 89; NJSFWC, 89–90; women's suffrage rights, 90; as writer, 90–91
Stevens, Edwin A., 77
stiffening trusses, 140, 182, 183, 189, 191, 213–14
stiff-leg derricks, 184, 185
Stockton, John P., 31
Storm King Highway, 94–95
Stryker, William B., 31–32
Stuyvesant, Peter, 20
Sunderland, Charles, 232
surety bonds, 162
suspender ropes/cables, 168, 174, 182, 190, 231
suspenders, 140, 180, 187, 213; deck, 174
suspension bridges, 8, 26; diagram of, 253. *See also* Bear Mountain Bridge (BMB)
suspension span, 188, 214, 230
Sutro, Frederick C., 228
Swiss Bureau of Insurance, 56

Tappan tribe, 20
Tappan Zee Bay, 103
Tappan Zee Bridge, 225
Tench, Ellen Murray, 147
Tench, Emma Harriet Roberts, 150–51

Tench, Frederick, 2–4, 9, 13, 14–15, 95, 141–43, 145–47, 221; Baird and, 137–38; BMB project, 127, 128, 133, 134, 157; children, 151; death and homage, 219; Emma and, 150–51; family business, 147–48; Harrimans and, 128, 133, 157; marriage, 151; Partridge and, 133; Perkins and, 159–60, 209–10; Roland and, 128, 133, 138, 159, 161–62; shipbuilding enterprise, 132; tunneling machine, 155; Union Bridge Company, 148; Union Pacific, 149; vehicular crossing, 124. *See also* construction (BMB)
Tench, William Eastwood Carruthers, 147
Tench and Son, 147–48
Tench Construction Co., 219
Terry, Edward, 132, 136, 147, 148; death, 219; railroads, 148–49; Rochester Bridge Company, 148; tunneling machine, 155; Union Bridge Company, 148; Union Pacific, 148–49
Terry, Rebecca Allie, 149
Terry, Tench: and Proctor Tunneling Machine Company, 155
Terry & Tench Co., Inc., 16, 137–38, 145, 147, 171, 174, 178, 181–88, 208–10, 215, 219; complicated projects, 156; contracts, 150, 152, 156–57; establishment, 150; federal government and, 157; field-personnel changes, 186; influence, 152; Lincoln Memorial, 157; union and nonunion projects, 152. *See also* construction (BMB)
Terry Shipbuilding Corporation, 156–57
Thompson-Seaton, Ernest, 84
Throgs Neck Bridge, 225
Time, 158, 244

Tooker & Marsh, 108
Torrey, Ray, 112
Trail Conference, 111–12
trails/trail system, 111–13
transportation, 23–29, 110. *See also* railroads/railways
transverse bracing, 187, 230
Trautman, Ralph, 77, 84
tribes, 19
Triborough Bridge, 225
Trowbridge, August, 116
Trowbridge, Katherine. *See* Perkins, Katherine Trowbridge
Trustees of Scenic and Historic Places and Objects in the State of New York. *See* American Scenic and Historic Preservation Society
tunneling machine, 155
Turners Station, 26
Twombly, Hamilton, 93

Union Bridge Company, 148
Union Pacific Railroad, 26, 49, 50, 148–50, 158
United States Fish Commission, 29
US Fish and Wildlife Service, 29
US Geological Survey, 29

Vanderbilt, Cornelius, III, 162–63, 165–66
Vanuxem, L. C., 53
Vermilye, Ashbel Green, 81–82
Vermilye, Elizabeth Breeze, 5, 66, 68–70, 72, 74, 79–82, 84–85, 89, 228–29; Bergen County Historical Society, 89; "The Bible as Literature," 82; League for the Preservation of the Palisades, 79–82

Vermilye, Helen Lansing DeWitt, 82
Vermule, C. C., 78
Verrazzano, Giovanni da, 18–19
Verrazzano-Narrows Bridge, 225
viaducts, 188, 190
Voorhees, Foster M., 67, 69, 72–73
Voorhees, John J., 121

W. A. Harriman and Co., 9, 132, 158, 162
Walters, George, 135
War of 1812, 21
War of Independence, 21
Watch Tower, 228
Wear, Joseph W., 168–69
Welch, Ashbel, 106
Welch, Camille Beall, 106–7
Welch, Jessie, 107
Welch, John, 106
Welch, Priscilla Addams, 106
Welch, William, 4, 14, 123, 140–41, 174; family background, 106–7; Jolliffe on, 105; marriage, 107; National Parks Conference, 110–11; Perkins and, 105, 107–8, 114, 118, 120; PIPC and, 97, 105, 107–14, 118, 159, 231, 232, 240; projects, 105; retirement, 240; traffic, 125; US Army, 106–7, 114, 136
Welch, William A., Jr., 107
Werts, George, 31, 33–34, 58
West, Charles E., 72
Western Union, 174
W. F. Carey and Co., Inc., 186–87, 208, 215
Wheeler, W. H., 31
White, J. DuPratt, 15, 77, 88, 103, 104, 109, 209, 210, 228, 240; BMB project and, 172; Dawson and, 123;

Demarest and, 122; Emery and, 123; memo listing pressing matters, 97; PIPC, 96–97
Williams, Robert, 58
Williamsburg Bridge, 152, 173, 185, 212, 213, 216, 225, 231, 232
Willis, Nathaniel Parker, 27–28
Winton, Henry D., 33–34
Woman's Suffrage Party, 90–91
Woman Voter, 90–91
women, 122–23; camping permits, 109; citizenship, 90; right to vote, 90

Women's Rights Convention, 90
World War I, 94, 114–16, 122–23, 132, 156
World War II, 97, 115, 225, 240–41, 244
Wright, Wilbur, 93

Yale School of Forestry, 86
YMCA, 116–19
Yonkers (NY) Herald, 219
Yonkers (NY) Herald Statesman, 139–40

Barbara Hansen Cali is a retired management executive with more than thirty years in the commercial real-estate and construction industry, serving at various times as estimator, project manager, property manager, and director. While vice president of property management for the New Jersey Division of Reckson Associates (a tristate REIT, owner, developer, and manager of Class A office buildings presently known as RXR Realty Corp.), she sat on the board of BOMA-NJ (Building Owners and Managers Association) and mentored junior managers in the good stewardship of building maintenance and quality construction procedures.

She is a member of the Orange County Historical Society and has published several articles in the *Journal of the Orange County Historical Society* as well as in the *Westchester Historian*, the journal of the Westchester Historical Society.

A lifelong advocate for children, she sits on the boards of Prevent Child Abuse–New Jersey and Child Wellness Institute (ChildWIN), two organizations dedicated to ensuring that all children enjoy a happy childhood.

www.ingramcontent.com/pod-product-compliance
Lightning Source LLC
Chambersburg PA
CBHW030010121025
33827CB00003B/4